Hendrik Bartko
OBSERVATION OF GALACTIC SOURCES OF VHE $\gamma$-RAYS

Hendrik Bartko

# Observation of Galactic Sources of Very High Energy $\gamma$-Rays with the MAGIC Telescope

Dissertation an der Fakultät für Physik
der Ludwig–Maximilians–Universität
München

vorgelegt von
Hendrik Bartko
aus Zossen

München, den 29.12.2006
Erstgutachter: Prof. Kiesling
Zweitgutachter: Prof. Genzel
Tag der mündlichen Prüfung: 09.03.2007

Bibliografische Information Der Deutschen Bibliothek
Die Deutsche Bibliothek verzeichnet diese Publikation in der Deutschen
Nationalbibliografie; detaillierte bibliographische Daten sind im Internet über
http://dnb.ddb.de abrufbar.

© HARLAND media, Lichtenberg (Odw.) 2007
www.harland-media.de

Gedruckt auf alterungsbeständigem Papier nach ISO 9706 (säure-, holz- und chlorfrei).

Printed in Germany

ISBN 978-3-938363-05-8

# Abstract

One of the most important 'messengers' of many high energy phenomena in our universe are $\gamma$-rays. The detection of very high energy (VHE) cosmic $\gamma$-radiation by ground-based Cherenkov telescopes has opened a new window to the Universe, called $\gamma$-ray astronomy. It is a rapidly expanding field with a wealth of new results, particularly during the last two years, due to the high sensitivity of a new generation of instruments. The major scientific objective of $\gamma$-ray astronomy is the understanding of the production, acceleration and reaction mechanisms of very high energy particles in astronomical objects. This is tightly linked to the search for sources of the cosmic rays.

The MAGIC (Major Atmospheric Gamma-ray Imaging Cherenkov) telescope is one of the new generation of Imaging Air Cherenkov Telescopes (IACT) for VHE $\gamma$-ray astronomy. With its 17 m diameter mirror the MAGIC telescope is today the largest operating single-dish IACT. It is located on the Canary Island La Palma (28.8°N, 17.8°W, 2200 m asl.).

Recently, eight new galactic VHE $\gamma$-ray sources were detected by the HESS collaboration. They have either no or very weak counter-parts in other wavelengths. This makes them ideal candidates for accelerators of hadronic cosmic rays. The Galactic Center was also found to be a source of VHE $\gamma$-rays by various groups. However, the reported spectra differed significantly such that the nature of the source could not yet be identified.

In this thesis, observations of three galactic sources of VHE $\gamma$-rays with the MAGIC telescope are discussed: the source at the Galactic Center and two sources in the galactic disc HESS J1813-178 and HESS J1834-087. The positions, extensions, morphologies and the differential fluxes of these sources are presented using the data from the MAGIC telescope, and possible flux variations with time are studied. To identify the $\gamma$-ray production mechanism and the nature of the sources, the $\gamma$-ray sources are related to possible counter-parts in other wavelength bands. For HESS J1813-178, leptonic and hadronic models for the multiwavelength emission are developed and compared to the data to identify the physical processes at work in the source. The source at the Galactic Center is shown to be a stable emitter of VHE $\gamma$-rays, and the implications for the source models are discussed.

As these sources are located in the southern sky and can only be observed under large zenith angles with the MAGIC telescope, suitable observation and analysis procedures for large zenith angles had to be developed. In order to achieve the best possible background determination, the sources were observed in the off-source tracking observation mode.

To further increase the sensitivity of the MAGIC telescope, new ultra-fast read-out electronics components have been developed as an upgrade project for the MAGIC telescope. The performance of the new system is evaluated based on prototype tests in the MAGIC telescope at La Palma. The production, tests, and installation on the MAGIC telescope of a full-scale read-out system are described.

# Zusammenfassung

Die Beobachtung von sehr hochenergetischer kosmischer $\gamma$-Strahlung durch erdgebundene Cherenkov-Teleskope hat ein neues Fenster zu unserem Universum geöffnet: Die $\gamma$-Astronomie. Eine der bedeutendsten Fragestellungen der $\gamma$-Astronomie ist das Verständnis der Produktion, Beschleunigung und der Reaktionsmechanismen hochenergetischer Teilchen in astronomischen Objekten, die Suche nach den Beschleunigern der kosmischen Strahlung.

Das MAGIC (Major Atmospheric Gamma-ray Imaging Cherenkov) Teleskop ist ein abbildendes Cherenkov-Teleskop der neuesten Generation für die $\gamma$-Astronomie. Mit seinen 17 m Spiegeldurchmesser ist das MAGIC Teleskop das größte Cherenkov-Teleskop. Es befindet sich auf der Kanarischen Insel La Palma (28.8°N, 17.8°W, 2200 m über NN).

In 2005 wurde hochenergetische $\gamma$-Strahlung von acht neuen Quellen in der galaktischen Ebene von der HESS Kollaboration beobachtet. Diese Quellen haben entweder keine oder nur sehr schwache Emission in anderen Wellenlängen. Dadurch sind sie ideale Kandidaten für die Beschleuniger der kosmischen Strahlung. Auch aus Richtung des galaktischen Zentrums wurde hochenergetische $\gamma$-Strahlung von verschiedenen Gruppen beobachtet. Allerdings zeigten die veröffentlichten Spektren signifikante Unterschiede, so daß die Natur der Quelle nicht identifiziert werden konnte.

In dieser Arbeit werden die Beobachtungen von drei galaktischen Quellen hochenergetischer $\gamma$-Strahlung diskutiert: Die Quelle im galaktischen Zentrum, HESS J1813-178 und HESS J1834-087. Die Position, Ausdehnung, Morphologie und der differentielle $\gamma$-Fluß dieser Quellen wird präsentiert. Um den Mechanismus der $\gamma$-Strahlen Produktion und die Natur der Quelle zu identifizieren, werden die $\gamma$-Quellen mit Quellen in anderen Wellenlängenbereichen in Beziehung gesetzt. Für HESS J1813-178 werden leptonische und hadronische Modelle für die Emission über das gesamte Wellenlängenspektrum erstellt und mit den Daten verglichen. Ferner wird gezeigt, daß die $\gamma$-Strahlen Quelle im galaktischen Zentrum einen zeitlich konstanten Fluß von $\gamma$-Strahlen aussendet und die entsprechenden Auswirkungen auf die Modelle der Quelle diskutiert.

Da die beobachteten Quellen in der südlichen Hemisphäre des Himmels gelegen sind, können diese nur unter einem großen Zenitwinkel mit dem MAGIC Teleskop beobachtet werden. Deswegen werden in der Arbeit die entsprechenden Verfahren für die Beobachtung und Datenanalyse unter hohen Zenitwinkeln entwickelt. Um den Hintergrund best möglich bestimmen zu können, wurden die Quellen in einem speziellen Beobachtungsmodus etwas außerhalb des Kamerazentrums abgebildet.

Um die Sensitivität des MAGIC Teleskops noch weiter zu steigern, wurde eine neue, besonders schnelle Ausleseelektronik entwickelt. Die Leistung des neuen Elektroniksystems wurde in einem Prototypentest im MAGIC Teleskop untersucht. Die Produktion, Tests und die Installation des Gesamtsystems der neuen Elektronik wird beschrieben.

# Contents

| | |
|---|---|
| Acknowledgement | 1 |
| Introduction - This Thesis | 3 |
| **1 Very High Energy $\gamma$-Ray Astronomy** | **9** |
| 1.1 Observation of $\gamma$-Rays | 10 |
| 1.2 Production Mechanisms of VHE $\gamma$-Rays | 12 |
| 1.2.1 $\gamma$-Rays Produced in Hadronic Interactions | 12 |
| 1.2.2 Synchrotron and Inverse Compton Emission of VHE Electrons | 15 |
| 1.2.3 Non-thermal Bremsstrahlung of Electrons | 18 |
| 1.2.4 Curvature Radiation | 18 |
| 1.2.5 $\gamma$-Ray Production from Relic Particle Annihilation | 18 |
| 1.2.6 Relativistic Boosting of Lower Energy $\gamma$-Rays to Very High Energies | 19 |
| 1.3 Cosmic Rays | 19 |
| 1.4 Acceleration of Charged Particles to Very High Energies | 22 |
| 1.5 Galactic Sources of VHE $\gamma$-Rays | 23 |
| 1.5.1 The Milky Way | 24 |
| 1.5.2 Supernova Remnants | 25 |
| 1.5.3 Pulsars and Pulsar Wind Nebulae | 27 |
| 1.5.4 $\gamma$-Ray Binaries | 28 |
| 1.5.5 Giant Molecular Clouds | 30 |
| 1.5.6 The Galactic Center | 31 |
| 1.5.7 The Galaxy in the Light of VHE $\gamma$-Rays | 35 |
| 1.6 Extra Galactic Sources of VHE $\gamma$-Rays | 38 |
| 1.7 Searches for $\gamma$-Rays from Dark Matter Annihilation | 39 |
| 1.7.1 Possible Observation targets | 40 |
| 1.7.2 Expectations for MAGIC | 41 |
| 1.8 Choice of Observation Targets for this Thesis | 42 |
| **2 The MAGIC Telescope** | **45** |
| 2.1 The Imaging Air Cherenkov Technique | 46 |
| 2.1.1 Interactions of High Energy Particles within Air | 46 |
| 2.1.2 Development of Air Shower Cascades | 48 |
| 2.1.3 The Emission of Cherenkov Light | 51 |
| 2.1.4 Imaging of the Cherenkov Light from an Air Shower by a Telescope | 52 |

|  |  |  |  |
|---|---|---|---|
| 2.2 | The Hardware Layout of the MAGIC Telescope | | 52 |
| | 2.2.1 | Site Location | 55 |
| | 2.2.2 | Telescope Mechanics / The Drive System | 55 |
| | 2.2.3 | The Mirror System | 55 |
| | 2.2.4 | The MAGIC Camera | 56 |
| | 2.2.5 | The Trigger of MAGIC | 58 |
| | 2.2.6 | Data Acquisition and Signal Processing of the MAGIC Telescope | 59 |
| | 2.2.7 | The Calibration System | 60 |
| | 2.2.8 | Monte-Carlo Simulations | 60 |
| 2.3 | Observations with the MAGIC Telescope | | 62 |
| 2.4 | MAGIC II: The Second MAGIC Telescope | | 63 |

# 3 Data Analysis 65

|  |  |  |  |
|---|---|---|---|
| 3.1 | Charge/Arrival Time Extraction | | 66 |
| | 3.1.1 | Characteristics of the Current MAGIC Read-out System | 66 |
| | 3.1.2 | Pulse Shape Reconstruction | 67 |
| | 3.1.3 | Criteria for an Optimal Signal Extraction | 70 |
| | 3.1.4 | Signal Extraction Algorithms | 70 |
| | 3.1.5 | Monte Carlo Studies of Signal Extraction | 74 |
| | 3.1.6 | Pedestal Reconstruction | 77 |
| | 3.1.7 | Calibration Pulse Reconstruction | 80 |
| | 3.1.8 | Performance and Discussion | 80 |
| 3.2 | Event Reconstruction | | 81 |
| | 3.2.1 | Calibrations | 81 |
| | 3.2.2 | Bad Pixel Treatment | 85 |
| | 3.2.3 | Image Cleaning | 85 |
| | 3.2.4 | Image Parameterization / Hillas Parameters | 86 |
| | 3.2.5 | Gamma/Hadron Separation | 87 |
| | 3.2.6 | Energy Reconstruction | 90 |
| | 3.2.7 | Source Position Reconstruction: The Disp Method | 91 |
| 3.3 | $\gamma$-Ray Signal Reconstruction / Background Subtraction | | 93 |
| | 3.3.1 | Alpha Analysis | 94 |
| | 3.3.2 | Disp-Sky Map Analysis | 95 |
| 3.4 | Determination of the $\gamma$-Ray Energy Spectrum | | 101 |
| | 3.4.1 | Effective Collection Area | 102 |
| | 3.4.2 | Effective Observation Time | 103 |
| | 3.4.3 | Unfolding of the Energy Spectrum | 104 |
| 3.5 | Analysis of Data Taken in the Wobble Mode | | 106 |
| | 3.5.1 | Trigger Losses in Effective Area | 108 |
| | 3.5.2 | Definition of ON and OFF samples for the Wobble Mode, Gamma Signal Determination | 109 |
| | 3.5.3 | Comparison of the ON/OFF and Wobble Observation Modi | 113 |
| 3.6 | Basic Performance Parameters of the MAGIC Telescope | | 114 |
| | 3.6.1 | Sensitivity | 114 |
| | 3.6.2 | Angular Resolution | 115 |

## CONTENTS

|     | 3.6.3 | Energy Threshold / Energy Resolution | 115 |
| --- | --- | --- | --- |
|     | 3.6.4 | MC - Data Comparison | 116 |
|     | 3.6.5 | Observation at Large ZAs | 118 |
|     | 3.6.6 | Extended Sources | 120 |
| 3.7 | Systematic Errors | | 121 |

## 4 Observation of VHE $\gamma$-Rays from Galactic Sources — 125

- 4.1 Observation of VHE $\gamma$-Rays from the Galactic Center ............ 125
  - 4.1.1 Introduction ............................................. 125
  - 4.1.2 Observations with the MAGIC Telescope ................. 126
  - 4.1.3 Data Analysis .......................................... 127
  - 4.1.4 Discussion ............................................. 132
  - 4.1.5 Concluding Remarks .................................... 135
- 4.2 Observation of VHE $\gamma$-Rays from HESS J1813-178 .............. 137
  - 4.2.1 Introduction ............................................. 137
  - 4.2.2 Observations with the MAGIC Telescope ................. 137
  - 4.2.3 Data Analysis .......................................... 138
  - 4.2.4 Multiwavelength Source Modeling ....................... 141
  - 4.2.5 Concluding Remarks .................................... 146
- 4.3 Observation of VHE $\gamma$-Rays from HESS J1834-087 .............. 148
  - 4.3.1 Introduction ............................................. 148
  - 4.3.2 Observations with the MAGIC Telescope ................. 148
  - 4.3.3 Data Analysis .......................................... 149
  - 4.3.4 Discussion ............................................. 154
  - 4.3.5 Conclusions ............................................ 158

## 5 The Data Acquisition System Upgrade of MAGIC — 159

- 5.1 DAQ System Upgrade Considerations ........................... 160
- 5.2 The Ultra-fast Fiber-Optic MUX-FADC Data Acquisition System .... 161
  - 5.2.1 General MUX-Principle ................................. 162
  - 5.2.2 Optical Delays and Splitters ............................ 164
  - 5.2.3 MUX Electronics ....................................... 165
  - 5.2.4 FADC Read-Out ........................................ 168
- 5.3 Performance of the System Components ......................... 169
  - 5.3.1 Performance of the Optical Delays and Splitters .......... 170
  - 5.3.2 Performance of the MUX Electronics ..................... 171
  - 5.3.3 FADC Performance .................................... 172
- 5.4 Prototype Test in the MAGIC Telescope on La Palma ............. 174
  - 5.4.1 Setup of the Prototype Test ............................ 175
  - 5.4.2 The Data .............................................. 176
  - 5.4.3 Data analysis .......................................... 176
- 5.5 Discussion ................................................... 185
- 5.6 Production and Installation of the Full MUX-FADC System ....... 186

| | | |
|---|---|---|
| **6** | **Summary, Conclusion and Outlook** | **189** |
| | 6.1 The Galactic Center | 191 |
| | 6.2 HESS J1813-178 | 192 |
| | 6.3 HESS J1834-087 | 193 |
| | 6.4 Future Searches for the Accelerators of the Cosmic Rays | 193 |

| | |
|---|---|
| **List of Acronyms and Abbreviations** | **197** |
| **List of Figures** | **201** |
| **List of Tables** | **203** |
| **Bibliography** | **207** |
| **Curriculum Vitae** | **221** |

# Acknowledgement

As a member of the MAGIC collaboration, I am grateful to each of the almost 150 collaborators without whose time and effort, in both large and small ways, this work would not have been possible. In particular, I would like to thank the members of the MAGIC group at the Max-Planck-Institut für Physik (Werner-Heisenberg-Institut) in Munich for their constant support during my time as a Ph.D. student.

I would like to thank Prof. M. Teshima, for giving me the opportunity to do research at the Max-Planck-Institut für Physik and Prof. Ch. Kiesling for being my academic advisor at the Ludwig-Maximilian-Universität München. I am also extremely grateful to our MAGIC group leader, Dr. R. Mirzoyan, for taking such good care of me.

I am deeply obliged to Dr. W. Bednarek, Dr. R. Bock, Dr. F. Goebel, Dr. E. Lorenz, Dr. D. Torres and Dr. W. Wittek, with whom I shared long hours of work and discussion. It would not have been possible without their tireless efforts and skillful teaching. In particular, I would like to thank them for taking the time to read the drafts of this thesis and to thoroughly answer all my questions. D. Mazin and N. Otte are the best office mates I could imagine.

Finally, I would like to thank my parents for all the support and love they gave me.

# Introduction - This Thesis

Astronomical observations have been made since ancient times with the unaided human eye, having their origins in the religious practices of pre-history. However, visible light constitutes only a very narrow part of the entire electromagnetic spectrum. Since the middle of the 20th century, technical progress permitted the extension of observations of celestial objects to other wavelengths, invisible to the human eye. These are the radio ($< 10^{-3}$ eV), infrared ($O(0.1\,\mathrm{eV})$), ultraviolet ($O(10\,\mathrm{eV})$), X-ray ($O(\mathrm{keV})$) and $\gamma$-ray ($>$ 511 keV) wavelength bands. Their observation created entire new branches of science, which are still in rapid development.

Not a single celestial object is hot enough to thermally emit very high energy (VHE) $\gamma$-rays ($E_\gamma > 100$ GeV). These must be produced in extreme dynamical processes. As of today one has measured VHE $\gamma$-ray signals from active galactic nuclei (supermassive black holes in the center of galaxies), supernova remnants (exploded stars), pulsar wind-driven nebulae, and $\gamma$-ray binaries (binaries of a solar mass compact object and a giant star).

The earth is exposed to a continuous flux of high energy particles from space. This radiation mainly consists of positively charged nuclei and a few electrons, positrons, photons and an unknown number of neutrinos. Although the cosmic radiation has been known and intensively studied since its discovery by V. Hess (1912), its sources have not been unambiguously identified yet. The main difficulty in the search for the sources and the acceleration mechanisms of the cosmic rays is due to the diffusion of charged particles in the non-regular interstellar magnetic fields. Because of these fields, the charged cosmic rays lose completely their directional information on their way from the source to earth. In inelastic collisions of high energy cosmic rays with ambient matter $\gamma$-rays and neutrinos are produced. These neutral particles give direct information about their source, as their trajectories are not affected by magnetic fields. Due to their extremely small interaction cross-section neutrinos are very difficult to detect.

Cosmic ray particles with energies between 1 GeV and at least up to the knee feature of the energy spectrum at about $10^{15}$ eV are supposed to be accelerated in our galaxy (Hillas 2005). The flux of cosmic rays from other galaxies is expected to be negligible at that energies due to their large distance. For a long time shocks produced at supernova explosions have been considered as best candidates for the sources of the galactic component of the cosmic ray flux, see e.g. Baade & Zwicky (1934). To study the acceleration sites and the propagation of the charged cosmic rays in our galaxy one has to observe our galaxy in the light of VHE $\gamma$-rays.

Nevertheless, not all VHE $\gamma$-rays from galactic sources are due to the interactions of cosmic rays with ambient matter. There are also other mechanisms for the production of

VHE γ-rays like the inverse Compton scattering of ambient low energy photons by VHE electrons. In order to determine, which of the VHE γ-ray sources is also a source of the hadronic cosmic rays, for each individual source of VHE γ-rays, the physical processes of particle acceleration and γ-ray emission in this source have to be determined. A powerful tool is the modelling of the multiwavelength emission of the source and comparison to multiwavelength data.

A specially interesting region for the cosmic ray acceleration is the very center of our own galaxy, the Milky Way. The region around the Galactic Center contains most likely a super massive black hole of about $(3-4) \cdot 10^6$ solar masses (Schödel et al. 2002; Genzel et al. 2003; Eisenhauer 2005), supernova remnants, a pulsar wind nebula candidate, hot gas, and large magnetic fields between 10 $\mu$G up to a 1 mG (Uchida & Guesten 1995; Morris & Serabyn 1996; LaRosa et al. 2005). Recently, evidence for VHE γ-radiation from the Galactic Center was reported by the CANGAROO (Collaboration of Australia and Nippon for a GAmma Ray Observatory in the Outback, Tsuchiya et al. (2004)), VERITAS (Very Energetic Radiation Imaging Telescope Array System, Kosack et al. (2004)) and HESS (High Energy Stereoscopic System, Aharonian et al. (2004b)) collaborations. The fact that the measured spectra of the three groups differ significantly has stimulated discussions about the origin of the differences. The source of the VHE γ-radiation is still unclear.

The HESS collaboration has reported the detection of γ-ray emission above a few hundred GeV from eight new sources located in the Galactic Plane (Aharonian et al. 2005a), close to the Galactic Center. Most of these newly discovered sources are part of a new population of galactic VHE γ-ray sources. None of them had previously been predicted to be observable in the VHE γ-ray domain. They also have only very weak counter-parts in other wavelengths – or none at all.

In my thesis I want to determine the nature and the physical properties of the VHE γ-ray source at the Galactic Center and some of the newly detected VHE γ-ray sources in the Galactic Plane. For this goal, it is important to exactly measure the location, extension, energy spectrum and source variability in γ-rays. Moreover, to identify the γ-ray production mechanism and the nature of the source, I relate the γ-ray source to possible counter-parts in other wavelength bands, located positionally coincident in the sky to the VHE γ-ray source. Only the spectral modelling of the multiwavelength emission can uniquely identify the physical processes at work in the source.

The earth's atmosphere corresponds to a "calorimeter" with a thickness of about 28 radiation length. Therefore, it is not transparent for γ-rays, nor for X-rays. Consequently, instruments to measure these wavelengths are installed in satellites. Due to the limited size of the detectors and the exponential drop of the cosmic ray spectrum with increasing energy, the satellite borne γ-ray telescopes have a very limited photon statistics for energies above a few tens of GeV. Future projects will push this limit towards higher values, though (Wood et al. 1995).

The development of high performance imaging air Cherenkov telescopes (IACTs) in the last 20 years allowed a comparably inexpensive ground-based measurement of very high energy γ-rays in the energy range above 300 GeV. In order to detect the primary γ-rays, an IACT produces images of the γ-ray induced air showers in the atmosphere using their emitted Cherenkov light.

Introduction - This Thesis 5

The MAGIC telescope is a new generation IACT, located on the Canary Island of La Palma (28.8°N, 17.8°W, 2200 m above sea level). With a mirror diameter of 17 m it is the largest single-dish IACT currently in operation. The main aim of MAGIC is to cover the unexplored energy range between the satellite based $\gamma$-ray telescopes and previous generation IACTs, i.e. the energy range between few tens of GeV up to about 300 GeV.

The observations of the Galactic Center and the new HESS sources in the inner galaxy (i.e. located close to Galactic Center on the sky) have to be conducted with the MAGIC telescope under large zenith angles (ZA) – up to 62° – due to the location of the sources on the southern sky. This large zenith angle, together with the location of the sources in bright star fields, require a careful choice of the observation modes and a dedicated analysis procedure, which will be developed in this thesis.

The MAGIC telescope does not only record $\gamma$-ray shower images, but it is also triggered by cosmic ray showers, single isolated muons and fluctuations from the light of the night sky. In fact, the background images are by a factor of up to several thousands more numerous than the images of $\gamma$-ray showers. Thus a statistical method has to be applied for the sample separation of $\gamma$-ray candidates (signal) and background events. It exploits the physical differences between hadronic and electromagnetic showers. In general, hadronic showers are broader, more irregular and subject to larger fluctuations. Also the arrival time structure of the Cherenkov light from $\gamma$-ray and hadron induced showers as well as single isolated muons shows differences.

The camera of the MAGIC telescope consists of 576 photomultiplier tubes (PMTs), which deliver about 2 ns full width at half maximum (FWHM) fast pulses from the $\gamma$-ray air shower Cherenkov light to the experimental control house. There they are passed through a 300 MSamples/s "flash" analog-to-digital converters (FADC) system. To record the pulse shape in detail, an artificial pulse stretching to about 6.5 ns FWHM is needed.

In this thesis I develop an FADC pulse reconstruction algorithm for Cherenkov telescopes and implement it in the common MAGIC analysis software. The algorithm computes the signal charge and the arrival time from the recorded FADC samples in each camera channel for each triggered Cherenkov light pulse. Based on the features of the read-out electronics, the reconstruction algorithm performs a numeric fit of the known signal shape to the recorded signal samples. The full noise autocorrelation will be taken into account.

Although quite satisfactory for many new measurements (see e.g. this thesis), the performance of the current read-out electronics is limited by the relatively slow sampling rate of 300 MSamples/s and the 6.5 ns pulse stretching. This causes a 'wash-out' of the pulse shape differences between hadron and $\gamma$-ray induced showers as well as a rather large integrated noise due to the light of the night sky. For the fast Cherenkov pulses an FADC with 2 GSamples/s can provide an improved reconstruction of the pulse shape, which should improve the $\gamma$/hadron separation, minimizing at the same time the integrated noise.

Fast FADCs with GSamples/s are available commercially; they are, however, very expensive and power consuming. The aim for the hardware part for my thesis work is to develop a fiber-optic multiplexing technique which uses a single 2 GSamples/s FADC to digitize 16 read-out channels consecutively. This multiplexed FADC read-out will greatly reduce the cost compared to using one ultra-fast FADC per read-out channel and will permit the production of a full scale ultra-fast read-out system as an upgrade of the MAGIC telescope.

Thus, my thesis has the following four main aims.

1. Observation of very high energy $\gamma$-rays from galactic sources with the MAGIC telescope. Modelling and discussion of the multi-wavelength emission of the sources.

2. Development of dedicated observation and analysis procedures for large zenith angles. Implementation of the off-source tracking observation mode (the so-called wobble observation mode). Computation of VHE $\gamma$-ray sky maps and comparison with multiwavelength data.

3. Development and implementation of an FADC pulse reconstruction algorithm in the common MAGIC analysis software framework MARS.

4. Development of new ultra-fast read-out electronics for the upgrade of the MAGIC telescope, including prototype-tests, production and installation of the full-scale system.

This thesis is organized in five chapters.

1. A short introduction into **VHE $\gamma$-ray astronomy**: The search for sources of VHE $\gamma$-rays is linked to the search for the sources of the cosmic rays. The known sources of VHE $\gamma$-rays in our galaxy and beyond are discussed. Possible $\gamma$-ray production mechanisms and the corresponding models for the multiwavelength source emission are studied. Finally, the choice of the observation targets of this thesis is explained.

2. **The MAGIC Telescope**: The technique of imaging air Cherenkov telescopes is described in general, including the production of Cherenkov light in air shower cascades and its imaging by a telescope. The particular hardware layout of the MAGIC telescope and its observations are presented in detail.

3. **Data Analysis**: To analyze the data of the MAGIC telescope an FADC signal reconstruction algorithm is developed. The calibration, image parameterization, $\gamma$/hadron separation, energy and arrival direction reconstruction are discussed. Subsequently, $\gamma$-ray sky maps and spectra are calculated. Dedicated reconstruction algorithms for data taken in the wobble observation mode are developed and the special requirements for large zenith angle observations are discussed. The basic performance parameters of the MAGIC telescope, including the systematic errors, are investigated.

4. **Observations of VHE $\gamma$-rays from Galactic Sources** are discussed: the Galactic Center, HESS J1813-178 and HESS J 1834-087. The source positions, extensions and the energy spectrum of the VHE $\gamma$-rays are determined and possible VHE $\gamma$-ray flux variations with time are studied. The results are put in the perspective of multiwavelength observations and the models of the multiwavelength emission.

5. **The Data Acquisition System Upgrade of the MAGIC Telescope**: In order to increase the sensitivity of the MAGIC telescope for future observations, a new ultra-fast data acquisition system using the novel technique of fiber-optic multiplexing is developed. The performance of the new data acquisition system is accessed in a prototype test in the MAGIC telescope. The full system is build, its quality controlled, installed and commissioned in the MAGIC telescope.

6. **Conclusion and Outlook**: The results of this thesis work are summarized and based on this knowledge an outlook to future observations and developments is given.

# Chapter 1
# Very High Energy $\gamma$-Ray Astronomy

Astronomy (Greek: $\alpha\sigma\tau\rho o\nu o\mu\iota\alpha = \alpha\sigma\tau\rho o\nu + \nu o\mu o\varsigma$, astronomia = astron + nomos, literally, "law of the stars") is the science of celestial objects (e.g. planets, stars and galaxies) and phenomena that originate outside the Earth's atmosphere (e.g. supernova explosions and the cosmic background radiation). The different disciplines of the observational astronomy can be defined by the type of the 'messenger' particle used (e.g. photons or, in future, possibly also neutrinos) and its energy. $\gamma$-ray astronomy uses photons with energies above the rest energy of an electron (511 keV). $\gamma$-rays with energies above 100 GeV are called Very High Energy (VHE) $\gamma$-rays. Part of the $\gamma$-ray astronomy is devoted to study the individual and collective properties of the astrophysical sources of VHE $\gamma$-rays such as

- pulsar wind nebulae
- shell-type supernova remnants
- $\gamma$-ray binary systems
- active galactic nuclei
- (still) unidentified VHE $\gamma$-ray sources.

The observation of VHE $\gamma$-rays can be used to answer fundamental questions in physics such as

- the origin of cosmic rays
- the so-called $\gamma$-ray horizon, which is directly related to the spectrum of the extragalactic background light (EBL)
- phenomena around compact objects (black holes and neutron stars)
- the probing of possible quantum gravity effects
- the nature of hypothetic Dark Matter particles.

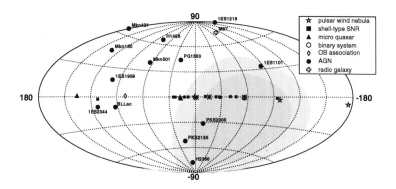

Figure 1.1: *The position of the 35 currently known VHE $\gamma$-ray sources in the sky (galactic coordinates). The white area is accessible to the MAGIC telescope. The light shaded area displays the sky region which the MAGIC telescope can observe only under large zenith angles between 50 and 70 degrees. The dark shaded area corresponds to even larger zenith angles and cannot be observed by the MAGIC telescope at all.*

Figure 1.1 shows the full sky in galactic coordinates indicating the currently known 35 VHE $\gamma$-ray sources. One can clearly distinguish two source populations, the sources clustering along the Galactic Plane as well as the uniformly distributed extra-galactic sources (AGNs from the blazar type and a radio galaxy).

This chapter is structured as follows: First, in section 1.1 the different instruments to observe $\gamma$-rays and their accessible energy bands are reviewed. Thereafter, the production mechanisms for $\gamma$-rays are modeled (section 1.2) and connected to the acceleration of charged particles to very high energies (section 1.3) and the production of the cosmic rays (section 1.4). In sections 1.5 and 1.6 the different types of galactic and extra-galactic sources of VHE $\gamma$-rays are introduced. Section 1.7 studies the prospects to search for $\gamma$-rays from hypothetical Dark Matter particle annihilation with the MAGIC telescope. Finally, based on the presented status of the field of $\gamma$-ray astronomy, in section 1.8 the targets for observation with the MAGIC telescope are chosen for this thesis.

## 1.1 Observation of $\gamma$-Rays

$\gamma$-rays have been detected within eight decades of energy, between 1 MeV and about 100 TeV (Aharonian et al. 2004a). They are expected to exist also up to the highest measured particle energies of above $10^{20}$ eV. This large energy range of $\gamma$-rays is covered rather inhomogeneously by completely different detection methods and flux sensitivities:

Soft $\gamma$-rays up to about 10 MeV can be observed by so-called coded mask instruments like the ones on board the INTEGRAL satellite (INTErnational Gamma-Ray Astrophysics

## 1.1 Observation of γ-Rays

Laboratory, Winkler et al. (2003)) or by scintillation detectors like OSSE (the Oriented Scintillation Spectrometer Experiment, Johnson et al. (1993)). Above that energy, instruments, which utilize Compton scattering of the γ-rays in the detector, offer the highest sensitivity. Examples are COMPTEL (the Compton Telescope, Diehl (1988)) on board the CGRO (Compton Gamma Ray Observatory) satellite and the future MEGA (Medium Energy Gamma-Ray Astronomy, Ryan et al. (2004)) satellite project.

High energy γ-rays between 100 MeV and 100 GeV can be efficiently detected by space-borne pair-conversion telescopes. These instruments use the conversion of the γ-ray into an electron-positron pair in the detector material and measure the energy and direction of these charged particles. Examples are the EGRET (Energetic Gamma Ray Experiment Telescope, Kanbach et al. (1988)) and the future GLAST (Gamma-ray Large Area Space Telescope, Wood et al. (1995)). In general, the γ-ray flux decreases steeply with energy. Beyond energies of about 100 GeV the space-borne γ-ray telescopes suffer from very limited γ-ray statistics due to their small effective area (in the order of m$^2$).

The energy range above 100 GeV can be best observed with ground-based instruments which offer large effective areas. Up to several tens of TeV, Imaging Air Cherenkov Telescopes (IACTs) offer the best sensitivity. Historically the first source of VHE γ-rays, the Crab Nebula, was detected with high significance by the Whipple telescope in 1989 (Weekes et al. 1989). For energies above ten TeV air shower arrays, integrated air Cherenkov arrays like AIROBICC (AIr-shower Observation By Angle Integrating Cherenkov Counters, Karle et al. (1995)) and water Cherenkov telescopes like Milagro (see e.g. Shoup et al. (1994); Atkins et al. (2004)) offer good sensitivity.

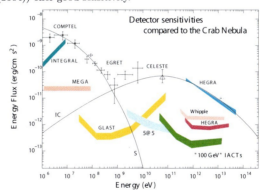

Figure 1.2: *Sensitivities of the different instruments (in operation or planned) to detect γ-rays in comparison to the energy spectrum of the Crab Nebula. The full line shows a model of the Crab Nebula emission. The green band labelled "100 GeV" IACTs shows the average sensitivity of the new generation IACTs CANGAROO, HESS, MAGIC, VERITAS. Figure from Aharonian (2004).*

Figure 1.2 shows the flux sensitivities of different γ-ray detection instruments, including future satellite missions like GLAST and MEGA, as well as high-altitude ground based telescopes (here the study 5@5 (Aharonian et al. 2001) is shown) in comparison to the predicted and detected flux from the Crab Nebula (figure taken from Aharonian (2004)).

## 1.2 Production Mechanisms of VHE $\gamma$-Rays

According to Planck's law, the average energy of the thermal black body radiation is directly linked to its temperature. Stars with typical surface temperatures of around 6000 K (like our sun) emit visible light with a tail extending to X-ray energies. The hottest observed objects in the universe like accretion discs around compact objects emit X-rays in the range of up to tens of keV. Nevertheless, no celestial object is hot enough to emit photons in the VHE $\gamma$-ray range. Therefore, the $\gamma$-rays must be produced in non-thermal extreme dynamic processes, which will be discussed in this section:

- decays of neutral mesons produced in interactions of VHE hadrons, section 1.2.1
- inverse Compton scattering of ambient low energy photons by VHE electrons, section 1.2.2
- bremsstrahlung of VHE electrons in ambient matter, section 1.2.3
- curvature radiation, section 1.2.4
- decays of hypothetic heavy relic particles from the big bang, section 1.2.5
- relativistic boosting of lower energy $\gamma$-rays to VHE, section 1.2.6.

In optically dense sources, internal absorption of $\gamma$-rays due to electron positron pair production in $\gamma\gamma$ collisions has to be taken into account, see e.g. Gould & Schreder (1966); Biller et al. (1995). Nevertheless, this process will not be discussed in this thesis.

### 1.2.1 $\gamma$-Rays Produced in Hadronic Interactions

Most of the VHE cosmic rays observed on earth are protons and heavier nuclei, see section 1.4. These particles produce VHE $\gamma$-rays in inelastic interactions with ambient matter via production and subsequent decay of secondary pions, kaons, hyperons, etc. The most important process is the decay of $\pi^0$ mesons into two $\gamma$-rays:

$$p + p \to \pi^0 + X \to \gamma\gamma + X \ . \tag{1.1}$$

Also the production and subsequent decay of other neutral mesons like $\eta$ contributes a few percent to the $\gamma$-ray signal. In addition, protons (and hadrons in general) can also produce $\pi^0$s via the process of photo-meson production: $p + \gamma \to p + \pi^0$. The threshold center of mass system energy for this process is the sum of the proton and $\pi^0$ masses. Therefore, this process requires either extremely high proton energies above $10^{19}$ eV in the case of proton scattering of the Cosmic Microwave Background (CMB) (Greisen 1966;

## 1.2 Production Mechanisms of VHE $\gamma$-Rays

Zatsepin & Kuzmin 1966) or higher energy photon fields (e.g. visible light), which require correspondingly lower proton energies of around $10^{15}$ eV.

The observation of $\pi^0$ decay $\gamma$-rays near the acceleration site of the hadronic cosmic rays offers the opportunity to probe the acceleration mechanism of the cosmic rays, see e.g. Ginzburg & Syrovatskii (1964) and Aharonian (2004). In addition to $\pi^0$s, also charged pions are produced in similar numbers in the proton proton collisions. These charged pions subsequently decay into muons and electrons and the corresponding (anti-) neutrinos. As an example the production and decay of a positive pion proceeds according to the following reaction chain:

$$p + p \rightarrow \pi^+ + X \rightarrow \mu^+ + \nu_\mu + X \rightarrow e^+ + \nu_e + \nu_\mu + \overline{\nu_\mu} + X \ . \tag{1.2}$$

The production of charged and neutral pions in hadronic collisions provides an important link between the $\gamma$-ray and future neutrino astronomy (for a review see e.g. Halzen & Hooper (2002)). The observation of neutrinos in the future would prove the hadronic production of the $\gamma$-rays observed today. Unfortunately, the neutrino interaction cross-section is minute, which requires km$^3$ detectors to observe a significant neutrino signal from galactic $\gamma$-ray sources, see e.g. Kappes et al. (2006).

In the following the $\gamma$-ray spectrum produced by a population of cosmic ray protons confined in a source is derived: Let $N_p(E_p)$ be the number of protons with minimum energy $E_p$, then the flux of protons in the energy interval $[E_p, E_p + \mathrm{d}E_p]$ is $J_p(E_p) = c\beta/(4\pi)\mathrm{d}N_p(E_p)/(\mathrm{d}E_p \mathrm{d}V)$. These high-energy protons react with ambient protons and nuclei of number density $n$ (for example in a molecular cloud or from the wind of a star) and may produce $\pi^0$s. The number of $\pi^0$s in the energy interval $[E_p, E_p + \mathrm{d}E_p]$ per source volume (the so-called $\pi^0$ emissivity $Q_{\pi^0}(E_{\pi^0})$) is then obtained to be (see e.g. Stecker (1971)):

$$Q_{\pi^0}(E_{\pi^0}) = 4\pi n \int_{E_{\mathrm{th}}(E_{\pi^0})}^{E_p^{\mathrm{max}}} \mathrm{d}E_p \, J_p(E_p) \frac{\mathrm{d}\sigma(E_{\pi^0}, E_p)}{\mathrm{d}E_{\pi^0}} \ , \tag{1.3}$$

where $E_p^{\mathrm{max}}$ is the maximum energy of the protons in the system, and $E_{\mathrm{th}}(E_{\pi^0})$ is the minimum proton energy required to produce a pion with total energy $E_{\pi^0}$, which is determined by the reaction kinematics, see e.g. Moskalenko & Strong (1998) and Blatning et al. (2000). Finally, $\mathrm{d}\sigma(E_{\pi^0}, E_p)/\mathrm{d}E_{\pi^0}$ is the differential cross section for the production of a pion with energy $E_{\pi^0}$, in the lab frame, due to a collision of a proton of energy $E_p$ with a hydrogen atom basically at rest. Different parameterizations of cross-sections have been studied in depth by Domingo-Santamaria & Torres (2005): One possibility is the use of a parameterization of the differential cross-section like the one given in Blatning et al. (2000). Another way is to relate the differential cross-section for the $\pi^0$ production to the total cross-section of inelastic proton-proton collisions, $\sigma_{\mathrm{pp}}(E_p)$, the so-called $\delta$-functional approximation (Aharonian & Atoyan 2000):

$$\frac{\mathrm{d}\sigma(E_{\pi^0}, E_p)}{\mathrm{d}E_{\pi^0}} = \delta(E_{\pi^0} - \kappa E_{\mathrm{kin}}) \, \sigma_{\mathrm{pp}}(E_p) \ , \tag{1.4}$$

where $\kappa$ is the mean fraction of the kinetic energy $E_{\mathrm{kin}} = E_p - m_p c^2$ of the proton transferred to the secondary neutral mesons ($\pi^0$, $\eta$ etc.) per collision. In a broad region from GeV to

TeV energies, $\kappa \sim 0.17$ (Gaisser 1990). In this approximation the $\pi^0$ emissivity is given as:

$$\begin{aligned} Q_{\pi^0}(E_{\pi^0}) &= 4\pi n \int_{E_{\text{th}}(E_{\pi^0})} dE_p\, J_p(E_p)\, \delta(E_{\pi^0} - \kappa E_{\text{kin}})\, \sigma_{\text{pp}}(E_p)\,, \\ &= \frac{4\pi n}{\kappa} J_p\left(m_p c^2 + \frac{E_{\pi^0}}{\kappa}\right) \sigma_{\text{pp}}\left(m_p c^2 + \frac{E_{\pi^0}}{\kappa}\right). \end{aligned} \quad (1.5)$$

Therefore, a reasonably good knowledge of the total inelastic cross section is needed. Aharonian & Atoyan (2000) proposed that, since from the threshold at $E_{\text{kin}} \sim 0.3$ GeV the cross section appears to rise rapidly to about 30 mb at energies of about $E_{\text{kin}} \sim 2$ GeV, and since after that energy it increases only logarithmically, a sufficiently good approximation is to assume:

$$\begin{aligned} \sigma_{\text{pp}} &\sim 30\,(0.95 + 0.06\ln(E_{\text{kin}}/\text{GeV}))\,\text{mb} \quad \text{for} \quad E > 1\,\text{GeV} \\ &\sim 0 \quad \text{otherwise.} \end{aligned} \quad (1.6)$$

The produced $\pi^0$s subsequently decay instantaneously (half-life about $10^{-16}$s) with a 98.8% chance into two $\gamma$-rays. Let $\phi$ be the angle (in the $\pi^0$ rest-system) between the boost direction of this system and the flight direction of one of the $\gamma$-rays. Then the $\gamma$-rays have, in the system of the observer, the energies $E_\gamma = E_{\pi^0}/2 \pm 1/2\sqrt{E_{\pi^0}^2 - m_{\pi^0}^2}\cos\phi$. As the $\phi$ values of the produced $\gamma$-rays are uniformly distributed, the probability density to get a $\gamma$-ray with energy $E_\gamma$ from a $\pi^0$ with energy $E_{\pi^0}$ is $2/\sqrt{E_{\pi^0}^2 - m_{\pi^0}^2}$. The minimum pion energy required to produce a photon of energy $E_\gamma$ is $E_{\pi^0}^{\min}(E_\gamma) = E_\gamma + m_{\pi^0}^2 c^4/(4E_\gamma)$. Hence, the $\gamma$-ray emissivity $Q_\gamma(E_\gamma)_{\pi^0}$ is obtained from the neutral pion emissivity $Q_{\pi^0}$ as:

$$Q_\gamma(E_\gamma)_{\pi^0} = 2\int_{E_{\pi^0}^{\min}(E_\gamma)}^{E_{\pi^0}^{\max}(E_p^{\max})} dE_{\pi^0}\, \frac{Q_{\pi^0}(E_{\pi^0})}{\sqrt{E_{\pi^0}^2 - m_{\pi^0}^2 c^4}}\,, \quad (1.7)$$

here $E_{\pi^0}^{\max}(E_p^{\max})$ is the maximum pion energy that the population of protons can produce, determined from the kinematics.

Assuming a uniform cosmic ray density and a uniform matter density, the flux of $\gamma$-rays coming from an emission region of volume $V$ at a distance $D$ is than given by:

$$\frac{dN_\gamma}{dE_\gamma dA dt} = \frac{V}{4\pi D^2} Q_\gamma(E_\gamma)\,. \quad (1.8)$$

The total energy in cosmic rays in the source, $W_p(E_p > 1\,\text{GeV})$, is (assuming a uniform cosmic ray density):

$$W_p(E_p > 1\,\text{GeV}) = V \int_{1\,\text{GeV}}^{\infty} \frac{dN_p(E_p)}{dV dE_p} dE_p\,. \quad (1.9)$$

In case of supernova remnants this energy can be compared to the supernova explosion power of typically $10^{51}$ erg (for reviews see e.g. Woltjer (1972); Jones et al. (1998)). Acceleration efficiencies up to around 10% are expected (Berezhko & Völk 1997, 2000).

## 1.2 Production Mechanisms of VHE $\gamma$-Rays

There is also a useful direct relation between the total cosmic ray energy stored in the source and its $\gamma$-ray luminosity $L_\gamma$:

$$L_\gamma = c n \sigma_{pp} \kappa W_p \, . \tag{1.10}$$

The probability of an inelastic interaction of a high-energy proton with the ambient matter is $c n \sigma_{pp}$. Assuming a constant value of $\sigma_{pp} = 30$ mb one can define a characteristic cooling time of the high energy protons due to $\gamma$-ray production:

$$t_{pp \to \gamma} = (c n \sigma_{pp} \kappa)^{-1} \sim 6 \times 10^{15} \mathrm{s} \left( \frac{n}{1 \mathrm{cm}^3} \right)^{-1} . \tag{1.11}$$

As will be shown in section 1.4, the accelerated protons and heavier nuclei follow a power-law spectrum, see also Bell (1978). Assuming a maximum achievable energy $E_{\max}$, one often assumes the following spectrum of high energy protons:

$$\frac{\mathrm{d}N_p(E_p)}{\mathrm{d}V \mathrm{d}E_p} = A_p (E_p/\mathrm{GeV})^{-\alpha} \exp(-E/E_{\max}) \, , \tag{1.12}$$

where $A_p$ is a normalization constant for the number density of protons per energy interval.

To summarize, a simple hadronic model for the $\gamma$-ray emission, as will be used in section 4.2, has the following parameters:

1. $A_p$: normaliztion constant of the number density of protons per energy interval
2. $\alpha$: the slope of the energy spectrum of the protons
3. $E_{\max}$: the maximum energy of the protons
4. $n$: the number density of the ambient matter
5. $V$: the volume of the $\gamma$-ray emission region
6. $D$: the distance to the source.

### 1.2.2 Synchrotron and Inverse Compton Emission of VHE Electrons

Suppose that VHE electrons are captured in a region characterized by a magnetic field $B$ and an ambient photon field of energy density $w_{\mathrm{ph}}$. In case of low matter density the electrons only produce a low bremsstrahlung power. They mainly lose energy due to synchrotron radiation (in the radio to X-rays) and inverse Compton (IC) up-scattering of the ambient photons to GeV-TeV energies. The proton mass is about 2000 times larger than the electron mass und the cross-sections for synchrotron radiation and IC up-scattering depend strongly on the particle mass. Therefore, VHE protons emit only negligible power in synchrotron and IC radiation. Only for energies above $10^{18}$ eV proton synchrotron radiation will play an important role.

The photon energy density $w_{\rm ph}$ receives contributions by several radiation fields such as:

1. the 2.7 K cosmic microwave background radiation
2. the diffuse galactic dust far-infrared (FIR) and starlight near-infrared/optical backgrounds
3. possible intensive radiation fields of local origin, especially the synchrotron radiation emitted from the same electron population.

In most cases, for the production of VHE $\gamma$-rays in galactic sources the 2.7 K CMB is the dominant target field (Aharonian, Atoyan & Kifune 1997) for the VHE electrons. For higher target photon energies the scattering cross-section, as given by the Klein-Nishina formula, decreases. Only in very dense radiation environments near the Galactic Center do other photon fields also contribute (Aharonian et al. 2005b). In AGNs there is a very high density of synchrotron radiation such that this radiation is the dominant ambient photon field to be up-scattered, the so-called synchrotron-self-Compton (SSC) effect (Rees 1967; Urry & Padovani 1995).

The inverse Compton emissivity is given by (see e.g. Blumenthal & Gould (1970)):

$$Q_\gamma(E_\gamma)_{\rm IC} = \int_0^\infty n_{\rm ph}(\epsilon)\mathrm{d}\epsilon \int_{E_{\rm min}} \frac{\mathrm{d}\sigma(E_\gamma,\epsilon,E_e)}{\mathrm{d}E_\gamma} c \frac{\mathrm{d}N_e(E_e)}{\mathrm{d}V\mathrm{d}E_e} \mathrm{d}E_e \; , \qquad (1.13)$$

where $n_{\rm ph}(\epsilon)$ is the number density of the ambient low energy photons with energy in the interval $[\epsilon, \epsilon + \mathrm{d}\epsilon]$. $N_e(E_e)$ is the number of electrons with a minimum energy of $E_e$. $E_{\rm min}$ is the minimum electron energy needed to generate a photon of energy $E_\gamma$, i.e. $E_{\rm min} = (E_\gamma/2)\left[1 + (1 + (m_e c^2)^2/\epsilon E_\gamma)^{1/2}\right]$. $\mathrm{d}\sigma(E_\gamma,\epsilon,E_e)/\mathrm{d}E_\gamma$ is the Klein-Nishina differential cross-section for the inverse Compton scattering (see e.g. Schlickeiser (2002)):

$$\frac{\mathrm{d}\sigma(E_\gamma,\epsilon,E_e)}{\mathrm{d}E_\gamma} = \left[3\sigma_{\rm T}(m_e c^2)^2/4\epsilon E_e^2\right]\left[2q\ln 2 + (1+2q)(1-q) + \frac{(Cq)^2(1-q)}{2(1+Cq)}\right] \; , \qquad (1.14)$$

where $\sigma_{\rm T} = 6.65 \times 10^{-25}$ cm$^2$ is the Thomson cross-section, $C = 4\epsilon E_e/(m_e c^2)^2$ is the so-called Compton factor, and $q = E_\gamma/[C(E_e - E_\gamma)]$.

Assuming a uniform distribution of the VHE electrons and of the ambient photons, the flux of $\gamma$-rays coming from an emission region of volume $V$ at a distance $D$ is then given by:

$$\frac{\mathrm{d}N_\gamma}{\mathrm{d}E_\gamma \mathrm{d}A \mathrm{d}t} = \frac{V}{4\pi D^2} Q_\gamma(E_\gamma) \; . \qquad (1.15)$$

In addition to the $\gamma$-ray production via IC scattering, the VHE electrons also produce synchrotron radiation, see e.g. Jackson (1975), in the ambient magnetic fields (generally extending from radio to X-rays). The emitted synchrotron radiation energy of a relativistic electron or positron per unit time per unit frequency interval, as a function of frequency $\nu$, is given by (see e.g. Kembhave & Narlikar (1999)):

$$P(E_e,\nu) = \sqrt{3}(eB)\sin\phi F(\nu/\nu_{\rm c})(e^2/m_e c^2) \; . \qquad (1.16)$$

## 1.2 Production Mechanisms of VHE $\gamma$-Rays

Here, $\nu_c = 3eB\sin\phi(E_e/m_ec^2)^2/(4\pi m_e c)$ is the so-called critical frequency, $E_e$ and $m_e$ are the energy and the mass of the electron, $B$ is the absolute value of the magnetic field, $\phi$ is the magnetic field pitch angle, and $F(x) = x\int_x^\infty K_{5/3}(\xi)\mathrm{d}\xi$, with $K_{5/3}$ being the modified Bessel function of order 5/3. The power emitted by all electrons per frequency interval, $\epsilon_{\mathrm{Sync}}(\nu)$, can then be obtained by integrating $P(E_e,\nu)$ over the distribution of the electron energies and averaging of the magnetic field pitch angles:

$$\epsilon_{\mathrm{Sync}}(\nu) = \frac{\sqrt{3}(eB)e^2}{m_ec^2}\int \mathrm{d}E_e \frac{\mathrm{d}N_e(E_e)}{\mathrm{d}V\mathrm{d}E_e}\int_0^{\pi/2}\mathrm{d}\phi\frac{\nu}{\nu_c}\sin^2\phi\int_{\nu/\nu_c}^\infty \mathrm{d}\xi K_{5/3}(\xi) \ . \tag{1.17}$$

Assuming a uniform distribution of VHE electrons and a uniform magnetic field, the energy flux of the synchrotron photons coming from an emission region of volume $V_{\mathrm{sync}}$ at a distance $D$ is than given by:

$$E_{\mathrm{sync}}\frac{\mathrm{d}N_{\mathrm{sync}}}{\mathrm{d}E_{\mathrm{sync}}\mathrm{d}A\mathrm{d}t} = \frac{V_{\mathrm{sync}}}{4\pi D^2}\epsilon_{\mathrm{Sync}}\left(\frac{\nu}{h}\right) \ . \tag{1.18}$$

In principle only part of the IC emitting volume may be filled with the magnetic field, which is defined as the magnetic filling fraction $f_B$:

$$V_{\mathrm{sync}} = f_B V_{\mathrm{IC}} \ . \tag{1.19}$$

In most particle acceleration models the accelerated particles follow a power-law spectrum (see section 1.4 and Bell (1978)). Modelling the achievable maximum energy as an exponential cut-off, the electron distribution can be described by:

$$\frac{\mathrm{d}N_e(E_e)}{\mathrm{d}V\mathrm{d}E_e} = A_e(E/\mathrm{GeV})^{-\alpha}\exp\left(-E/E_{\mathrm{max}}\right)\mathrm{GeV}^{-1}\mathrm{cm}^{-3} \ , \tag{1.20}$$

where $A_e$ is a normalization constant. Thus a simple leptonic model for the multiwavelength radio to $\gamma$-ray emission, as used in section 4.2, uses the following seven parameters:

1. $A_e$: the number density of electrons per energy interval

2. $\alpha$: the slope of the energy spectrum of the electrons

3. $E_{\mathrm{max}}$: the maximum energy of the electrons

4. $V_{\mathrm{IC}}$: the volume of the $\gamma$-ray emission region

5. $B$: the average magnetic field strength

6. $f_B$: the magnetic filling factor

7. $D$: the distance to the source.

## 1.2.3 Non-thermal Bremsstrahlung of Electrons

VHE electrons can also produce VHE $\gamma$-rays in interactions with the ambient matter, through the process of bremsstrahlung, see e.g. Heitler (1954). The $\gamma$-ray emissivity is proportional to the density of the ambient matter. In most astrophysical sources, the photon density is typically many orders of magnitude higher than the matter density. Therefore, in general VHE electrons lose their energy much more efficiently by inverse Compton scattering and synchrotron radiation than by bremsstrahlung (see e.g. Aharonian (2004)). Only in very dense environments (like for example in the $\gamma$ Cygni SNR with $n = 300\,\text{cm}^{-3}$ (Uchiyama et al. 2002)) bremsstrahlung may dominate. The cross section for the bremsstrahlung process depends sensitively on the particle mass. Due to the high mass of protons compared to electrons, bremsstrahlung energy losses for protons can be neglected.

## 1.2.4 Curvature Radiation

Like synchrotron radiation, curvature radiation is caused by charged particles being accelerated in magnetic fields. However, in the case of curvature radiation the acceleration is due to the motion of the charged particle along the curved magnetic field lines, see e.g. Zhang & Yuan (1998). Magnetic field lines are often curved, e.g. in pulsar magnetospheres (Lovelace 1976; Bednarek 1997).

## 1.2.5 $\gamma$-Ray Production from Relic Particle Annihilation

In the Big Bang some hypothetical weakly interacting massive particles might have been produced, which might still be abundant in some regions of the universe, see e.g. Scherrer & Turner (1986); Jungman et al. (1996). One possibility for such particles is the neutralino $\chi$ (Ellis et al. 1984). Two neutralinos may annihilate with a rate $\Gamma$ per unit volume and time given by (Bergstrom et al. 1998):

$$\Gamma = \frac{1}{2} \cdot \frac{\langle \sigma v \rangle}{m_\chi^2} \rho_\chi^2 \,. \tag{1.21}$$

Here $m_\chi$ is the neutralino mass, $\rho_\chi$ is the neutralino density and $\langle \sigma v \rangle$ is the thermally averaged annihilation cross-section. The rate of $\gamma$-rays produced per energy interval in the source volume $\Delta V$ is then given by:

$$\frac{\mathrm{d}N_\gamma}{\mathrm{d}t\,\mathrm{d}E} = \frac{\mathrm{d}N_\gamma(m_\chi, \text{SUSY})}{\mathrm{d}E} \cdot \frac{1}{2} \cdot \frac{\langle \sigma v \rangle}{m_\chi^2} \int_{\Delta V} \rho_\chi^2(\vec{r})\mathrm{d}^3 r \,, \tag{1.22}$$

where $\frac{\mathrm{d}N_\gamma(m_\chi,\text{SUSY})}{\mathrm{d}E}$ is the so-called fragmentation function, i.e., the number of produced $\gamma$-rays per energy per neutralino annihilation, and $\vec{r}$ is the spatial position.

Let $d$ be the distance to the annihilation position, the $\gamma$-ray flux detectable on earth with an detector of active area $A$ is then given by:

$$\frac{\mathrm{d}N_\gamma}{\mathrm{d}t\,\mathrm{d}A\,\mathrm{d}E} = \frac{\mathrm{d}N_\gamma(m_\chi, \text{SUSY})}{\mathrm{d}E} \cdot \frac{1}{2} \cdot \frac{\langle \sigma v \rangle}{m_\chi^2} \frac{1}{4\pi d^2} \int_{\Delta V} \rho_\chi^2(\vec{r})\mathrm{d}^3 r \,. \tag{1.23}$$

Using the identity $\frac{1}{d^2}\int_{\Delta V}\rho_\chi^2(\vec{r})\mathrm{d}^3r = \int_{\Delta\Omega}\mathrm{d}\Omega\int_{\text{los}}\rho_\chi^2(\vec{r}(s,\Omega))\mathrm{d}s$, where $s$ is the distance on the line of sight (los) from the observer, the differential flux per solid angle can be obtained to be:

$$\frac{\mathrm{d}N_\gamma}{\mathrm{d}t\,\mathrm{d}A\,\mathrm{d}E} = \frac{\mathrm{d}N_\gamma(m_\chi,\text{SUSY})}{\mathrm{d}E}\cdot\frac{1}{2}\cdot\frac{\langle\sigma v\rangle}{4\pi m_\chi^2}\cdot\int_{\Delta\Omega(\text{PSF})}\mathrm{d}\Omega\int_{\text{los}}\rho_\chi^2(\vec{r}(s,\Omega))\mathrm{d}s\;. \quad (1.24)$$

The solid angle $\Omega$ has to be integrated at least over a range corresponding the instruments PSF.

Thus the rate of observable $\gamma$-rays is a function of the following unknowns:

1. $\rho_\chi(\vec{r})$: the spatial distribution of the neutralinos with respect to the observer

2. $m_\chi$: the neutralino mass

3. $\frac{\mathrm{d}N_\gamma(m_\chi,\text{SUSY})}{\mathrm{d}E}$: the $\gamma$-ray yield per annihilation, which depends on $m_\chi$ and also (weakly) on all other SUSY parameters.

The energy spectrum of the produced $\gamma$-radiation has a very characteristic feature with a cut-off at the mass of the Dark Matter particle. Moreover, the flux should be absolutely stable in time.

### 1.2.6 Relativistic Boosting of Lower Energy $\gamma$-Rays to Very High Energies

In some cases, the $\gamma$-radiation is produced in an object which moves with a bulk Lorentz factor $\gamma$ (corresponding to the velocity $\beta c$). Examples are the jets of AGNs and microquasars (see sections 1.5.4 and 1.6) and maybe GRBs. Then the $\gamma$-ray energy in the restframe of the observer is given by (Rees & Sciama 1966; Krawczynski et al. 2004):

$$E_{\text{obs}} = [\gamma(1-\beta\cos\theta)]^{-1}E_{\text{emission}}\;, \quad (1.25)$$

where $\theta$ is the angle between the velocity of the emission region and the line of sight in the observer frame.

## 1.3 Cosmic Rays

The earth is exposed to a continuous flux of high energy particles from space. The cosmic rays arrive isotropically from space and have an energy density of about 1 eV/cm$^3$. This radiation mainly consists of positively charged nuclei and a few electrons, positrons, photons and neutrinos (the flux of neutrinos has not been measured yet). Figure 1.3 displays the energy spectrum of the charged cosmic rays, according to Cronin, Gaisser & Swordy (1997). For energies above about 1 GeV the spectrum follows a power law $\mathrm{d}N/\mathrm{d}E \sim E^{-\alpha}$ with a spectral index $\alpha \approx 2.7$. At an energy of a few $10^{15}$ eV the spectral index steepens to $\alpha \approx 3.0$ (known as the "knee" of the cosmic ray spectrum). At an energy of around $10^{18}$ eV the spectral index gets again harder, which is dubbed to be the "ankle" of the cosmic ray spectrum. The energy spectrum extends to at least about $10^{20}$ eV, an energy entirely out of reach for ground-based accelerators. A center of mass energy of 10 TeV of colliding accelerator beams corresponds to the center of mass energy of the collision of a $5 \cdot 10^{16}$ eV cosmic ray proton with a proton at rest. The shape of the spectrum for energies above $10^{20}$ eV is currently under intense experimental study, see e.g. Kampert et al. (2006).

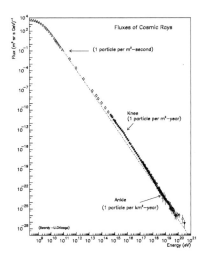

Figure 1.3: *Cosmic ray spectrum after Cronin, Gaisser & Swordy (1997). The "knee" and "ankle" features are indicated, see text for details.*

Although the cosmic radiation has been known and intensively studied since its discovery by V. Hess (1912), its sources have not been unambiguously identified yet (for a recent review, see e.g. Gaisser (2001); Yao et al. (2006)). The relevant questions are a) which astrophysical objects are the main sources of the cosmic rays, and b) which are the mechanisms for their acceleration.

The main difficulty in the search for the sources and the acceleration mechanisms of the cosmic rays is due to the diffusion of charged particles in the non-regular interstellar magnetic fields. The charged cosmic rays lose completely their directional information on their way from the source to earth. In addition, the spectrum of the cosmic rays at their acceleration site may be significantly altered by this diffusion process to yield the observed spectrum at earth. Thus, from charged cosmic rays the identification of the sources of the cosmic radiation is not possible. In contrast, $\gamma$-rays and neutrinos give direct information about their source, as their trajectories are not affected by magnetic fields.

Cosmic rays with an energy below 1 GeV are mainly produced in the sun as shown by several space missions. Particles with higher energy (at least up to the knee feature of the

## 1.3 Cosmic Rays

spectrum at about $10^{15}$ eV) are supposed to be accelerated in our own galaxy (Hillas 2005), while particles with energies above the "ankle" ($10^{18}$ eV) are assumed to be accelerated in extra-galactic sources. The origin of the particles with energies between the knee and the ankle is a matter of recent scientific discussion. Data may suggest that the spectral break at the knee is due to a limitation in the maximum energy of the cosmic ray accelerator (Haungs 2003) and the cosmic rays between the knee and the ankle are produced in the same galactic sources as the cosmic rays with energies below the knee.

For decades supernova remnants have been assumed to be sources of the galactic cosmic rays (Baade & Zwicky 1934; Ginzburg & Syrovatskii 1964). In the case this is true, supernova remnants should also be sources of high energy $\gamma$-rays, which are produced in the interactions of the accelerated particles (electrons and hadrons) with the ambient matter and radiation fields near the accelerator (Aharonian, Drury & Voelk 1994). This is the main motivation to study supernova remnants. Historically, the search for the sources of the cosmic rays was even one of the main motivations for the development of Cherenkov telescopes. Nevertheless, the VHE $\gamma$-rays may either be produced by interactions of VHE electrons (e.g. Inverse Compton Scattering of the CMB, see section 1.2.2) in the source or by the reaction of accelerated cosmic ray hadrons with ambient matter (see section 1.2.1).

For the search for the sources of the galactic cosmic rays the total acceleration power and the source spectrum of the cosmic rays are important parameters. Approximating the galaxy as a cylinder of radius $R = 15$ kpc and height $d = 200$ pc, one obtains an estimate for the volume of the galaxy:

$$V = \pi R^2 d \approx \pi (15 \text{kpc})^2 (200 \text{pc}) \approx 4 \cdot 10^{66} \text{cm}^3 \ . \tag{1.26}$$

The average time of a cosmic ray particle spent in the galaxy is $\tau_{\text{esc}} \approx 3 \cdot 10^6 \text{a} \approx 10^{14} \text{s}$ (Gaisser 2001). During that time only about 6% of the cosmic rays may interact with the interstellar matter. If one furthermore assumes that the energy density of cosmic rays in the galaxy is constant and equal to the locally measured energy density of $\rho_E \approx 1$ eV/cm$^3$, one obtains the total power needed to accelerate the cosmic rays:

$$P = \frac{\rho_E V}{\tau_{\text{esc}}} \approx 4 \cdot 10^{52} \text{eV} \text{s}^{-1} \approx 7 \cdot 10^{40} \text{erg} \text{s}^{-1} \ . \tag{1.27}$$

If one assumes that there is one SN explosion every thirty years in our galaxy, the average power released from all SN explosions is $10^{51}\text{erg}/30\text{years} \approx 10^{42}\text{erg}\,\text{s}^{-1}$ (Ginzburg & Syrovatskii 1964).

Moreover, the time $\tau_{\text{esc}}$, which the cosmic ray particles spend in the disk and the halo of our galaxy, is energy dependent. Therefore, the spectrum of the measured cosmic rays is different from the energy spectrum at the accelerator:

$$\left(\frac{dN}{dE}\right)_{\text{source}} \times \tau_{\text{esc}}(E) = \left(\frac{dN}{dE}\right)_{\text{observed}} \propto E^{-2.7} \ . \tag{1.28}$$

The energy dependence of the escape time $\tau_{\text{esc}}(E)$ may be measured by comparing the spectrum of a secondary cosmic ray nucleus (created through interaction of the primary cosmic rays with ambient matter) to that of a parent primary nucleus (accelerated in the source), e.g. boron to carbon. It was found to follow a power law (Garcia-Monoz et al. 1987; Swordy et al. 1990):

$$\tau_{\rm esc}(E) \propto E^{-0.6} \ . \tag{1.29}$$

Therefore, the source spectrum should follow a power law with spectral index of around $-2.1$:

$$\left(\frac{{\rm d}N}{{\rm d}E}\right)_{\rm source} \propto E^{-2.1} \ . \tag{1.30}$$

Nevertheless, other authors suggest different spectral indices of $\tau_{\rm esc}(E)$ as high as $-0.3$ (Seo & Ptuskin 1994) such that also source spectral indices of $-2.4$ may be possible.

## 1.4 Acceleration of Charged Particles to Very High Energies

The most widely studied mechanism for the acceleration of charged particles to very high energies is the so-called diffusive shock acceleration (for reviews see e.g. Blandford & Eichler (1987); Jones & Ellison (1991); Longair (1994) and Malkov & Drury (2001)). It is based on the first order acceleration mechanism proposed by Fermi (1949, 1954). A shock front is a surface of discontinuity across which there is a steady flow of mass, momentum and energy. The distance, over which the flow variables vary, the shock thickness, should be much smaller than the corresponding scales ahead of and behind the shock. Therefore, the overall flow pattern does not change substantially in the time it takes a fluid element to cross the shock (Blandford & Eichler 1987). An example for a shock front is the accelerated stellar material of a supernova explosion, which expands into the interstellar material, see section 1.5.2.

The general principle of the first order Fermi acceleration is illustrated in figure 1.4: A flow of charged particles (from left to right) impinges on the interstellar matter at rest and creates a shock, which travels with velocity $\beta_S c$ (typically around $10^4$ km/s). The matter up-stream of the shock (right part) is at rest, while the swept-up matter down-stream (left part) travels with velocity $\beta_P c$ from left to right. Treating the interstellar matter in the up- and down-stream regions as an ideal gas, the matter density of the swept-up gas (down-stream) is four times the one up-stream of the shock and $\beta_P = 3/4\beta_S$ (Blandford & Eichler 1987).

A relativistic particle can cross the shock front in either direction in case the thickness of the shock is smaller than the gyro-radius of the particle. Let us consider the case when a particle of energy $E_1$ crosses the shock from the up-stream to the down-stream region. In the rest frame of the downstream material the particle has the energy $E_2' > E_1$. It is elastically scattered by irregular magnetic fields such that its flight direction becomes isotropic in the down-stream rest-frame. Eventually it crosses the shock. In the restframe of the up-stream gas the particle now has an energy of $E_2 > E_2' > E_1$. The particle again scatters off some magnetic fields and will cross the shock multiple times until it leaves the

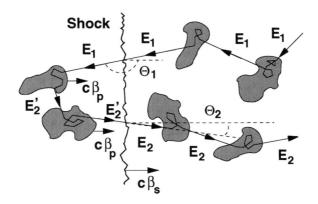

Figure 1.4: *Fermi acceleration: picture inspired from Longair (1994). See text.*

shock region by chance. On average the particle gains the energy $\Delta E$ per shock-crossing:

$$\frac{\Delta E}{E} \approx \beta_S \ . \tag{1.31}$$

An initially monoenergetic spectrum of relativistic particles evolves through this Fermi acceleration process into a power-law spectrum:

$$\frac{dN}{dE} \propto E^{-2-\epsilon} \ , \tag{1.32}$$

with $\epsilon$ being a small number of the order or 0.1, see e.g. Berezhko & Völk (1997, 2000). The particle acceleration may be a quite efficient process such that the influence of the accelerated particles on the shock has to be taken into account in non-linear calculations.

In addition to the diffusive shock acceleration, other mechanisms for the particle acceleration to very high energies are also discussed like acceleration in relativistic pulsar winds (Arons et al. 1998; Bednarek 2006), in the inner pulsar magnetosphere (Rudak 2001) or due to the reconnection of magnetic field lines near compact objects (Michel 1982; Coroniti 1990; Lyubarsky & Kirk 2001).

## 1.5 Galactic Sources of VHE $\gamma$-Rays

The previous chapters have linked the production of $\gamma$-rays to the presence of charged cosmic rays (either hadrons or leptons). In order to study the acceleration and propagation of the charged cosmic rays in the galaxy one has to study the galaxy in the light of VHE $\gamma$-rays. Therefore, first, some basic features of our galaxy, the Milky Way, will be reviewed in section 1.5.1. Thereafter, the known sources of VHE $\gamma$-rays like supernova remnants, pulsars and pulsar wind nebulae, compact binary systems, giant molecular clouds and the

Galactic Center will be described in sections 1.5.2 to 1.5.6. Finally, in section 1.5.7, the results of scans of the Galactic Plane in VHE $\gamma$-rays are presented.

## 1.5.1 The Milky Way

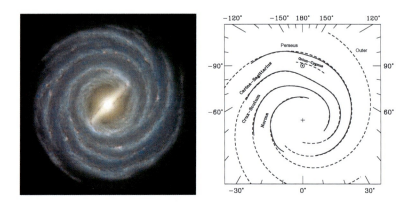

Figure 1.5: *The Structure of the Milky Way. Left: An artist's image displaying the Milky Way as a barred spiral galaxy, based on the results of the Spitzer infrared space telescope. Picture by Hurt (2005). Right: Overview of the location and naming of the spiral arms (Cordes & Lazio 2002). The Galactic Center is marked with a "+" sign and the sun is denoted by "⊙". The distance between the sun and the Galactic Center is about 8 kpc (Reid 1993). The angular scale shows the galactic longitude. The inner most region of the galaxy (where no spiral arms shown) contains the 3-5 kpc ring and a bar.*

The Milky Way (a translation of the Latin Via Lactea, in turn derived from the Greek γαλαξιας (galaxias)), is a barred spiral galaxy of Hubble type SBbc (loosely wound barred spiral). It is the home of the Earth and the Solar System. Figure 1.5 (left) shows an artist's image of the Milky Way based on the results of the Spitzer infrared space telescope, picture by Hurt (2005).

The term "milky" originates from the hazy band of white light appearing across the celestial sphere visible from Earth, which comprises stars and other material lying within the Galactic Plane. The galaxy appears brightest in the direction of Sagittarius, towards the Galactic Center, see also section 1.5.6 and section 4.1. The distance from the Sun to the Galactic Center is estimated to be about 8 kpc (Reid 1993). The Galactic Center harbors a compact object of very large mass, strongly suspected to be a supermassive black hole (Schödel et al. 2002; Ghez et al. 2003a). Most galaxies are believed to have a supermassive black hole at their center (Richstone et al. 1998; Gebhardt et al. 2000).

Relative to the celestial equator, the Milky Way passes as far north as the constellation of Cassiopeia and as far south as the constellation of Crux, indicating the high inclination

## 1.5 Galactic Sources of VHE γ-Rays

of Earth's equatorial plane and the plane of the ecliptic relative to the Galactic Plane. The fact that the Milky Way divides the night sky into two roughly equal hemispheres indicates that the solar system lies close to the Galactic Plane.

The main disk of the Milky Way Galaxy has a radius of about 15 kpc. Outside the Galactic core it has a thickness of about 200 pc. The disk bulges outward at the center. The Galactic Halo extends out to about 100 kpc in diameter. Our galaxy is composed of $(2-4) \cdot 10^{11}$ stars and has a total mass of $6 \cdot 10^{11}$ to $3 \cdot 10^{12}$ solar masses ($M_\odot$) (Battaglia et al. 2005). Most of the mass of the Milky Way is thought to be Dark Matter (see also section 1.7), forming a Dark Matter halo of an estimated $2 \cdot 10^{11}$ to $3 \cdot 10^{12}$ solar masses, which is concentrated towards the Galactic Center (Battaglia et al. 2005).

The galaxy's bar is thought to be about 8 kpc long, running through the center of the galaxy at an angle of about 45 degrees to the line between the sun and the center of the galaxy. It is composed primarily of red stars, believed to be ancient.

Each spiral arm describes a logarithmic spiral (as do the arms of all spiral galaxies) with a pitch of approximately 12 degrees. Four major spiral arms are believed to exist, which all start at the Galaxy's center, and at least two smaller arms or spurs. The right part of figure 1.5 names the different spiral arms (figure adapted from Taylor & Cordes (1993); Cordes & Lazio (2002)).

### 1.5.2 Supernova Remnants

In the following some important properties of supernovae and supernova remnants are reviewed, which are the basis for the discussion of the observation results of VHE γ-rays with the MAGIC telescope, presented in sections 4.1 to 4.3:

A supernova (SN) is either the runaway nuclear burning inside a matter-accreting white dwarf due to the external pressure (type Ia) or the catastrophic collapse of a massive star (types II or Ib/c), see e.g. Nomoto et al. (1984). In both cases an energy of about $10^{51}$ erg is released nearly instantaneously from the fusion of light elements in supernova type Ia or gravitational energy release in SN types II or Ib/c. For a review about SN mechanisms see e.g. Bethe (1990). The explosion energy accelerates the stellar material to speeds greater than the speed of sound (up to $10^4$ km/s) and causes a shock wave to move outwards from the central star. The high velocity stellar material plows outwards into the interstellar medium (ISM), compressing and heating ambient gas and sweeping it up much as a snow plow compacts and sweeps up snow. The ISM becomes enriched with the stellar material blown off in the explosion. The expanding material, and any additional material the blast collects as it travels through the interstellar medium, forms a supernova remnant (SNR). Hence, an SNR consists of a thin expanding shell with a relatively low density interior.

According to Woltjer (1972), one can distinguish four phases of the development of SNRs:

- **Free expansion phase**: As long as the material swept up by the shock is much less than the mass of the stellar ejecta, the expansion of the stellar ejecta proceeds at essentially a constant velocity equal to the initial shock wave speed, typically of the order of $10^4$ km/s. This may last for approximately 200 years, at which point the shock wave has swept up as much interstellar material as the initial stellar ejecta. At this time the SNR has on average a radius of 3 pc. Although the remnant is radiating thermal X-ray and synchrotron radiation (from radio to X-rays), the initial energy of the shock wave will have diminished very little. Line emission from the radioactive isotopes generated in the supernova contributes significantly to the total apparent brightness of the remnant in the early years, but do not significantly affect the shock wave.

- **Sedov-Taylor phase**: As the remnant sweeps up ambient mass about equal to the mass of the stellar ejecta, the wave will begin to slow down and the remnant enters a phase known as adiabatic expansion (Sedov-Taylor or blast wave phase). The internal energy of the shock continues to be very large compared to radiation losses from thermal and synchrotron radiation, so the total energy remains nearly constant. The rate of expansion is determined only by the initial energy of the shock wave and the density of the interstellar medium. This phase is believed to be the main phase of cosmic ray acceleration in the SNR.

- **Radiative phase**: As the shock wave cools, it will become more efficient to radiate energy. Once the temperature drops below about $2 \cdot 10^4$ K, electrons will be able to recombine with carbon and oxygen ions, enabling ultraviolet line emission which is a much more efficient radiation mechanism than the thermal X-ray and synchrotron radiation.

- **Dispersal phase**: X-radiation becomes much less apparent and the remnant cools and disperses into the surrounding medium over the course of the next $10^4$ years. Finally, the SNR has dispersed into the ISM and can hardly be recognized as an individual object.

Supernova remnants are extremely important for the understanding of our Galaxy. They provide the dominant energy input to the ISM. Therefore, they are believed to be responsible for the acceleration of galactic cosmic rays. Heavy elements (up to iron) created by fusion in the stellar core are released into the galaxy by the mixing of the ejecta and ISM material in the remnant. Elements heavier than iron are created in the powerful blast of a SN explosion, see e.g. Woosley & Weaver (1995).

For a long time Galactic SNRs have been considered as best candidates for the acceleration of hadronic cosmic rays up to the knee in the cosmic-ray particle spectrum at $10^{15}$ eV, see e.g. Baade & Zwicky (1934), Ginzburg & Syrovatskii (1964), Aharonian, Drury & Voelk (1994). For reviews see e.g. Hillas (2005) and Gaisser (2001). This hypothesis is based on three reasons:

- The total cosmic ray power equals about 10% of the average SN power in our galaxy, see section 1.3.

- supernova explosions produce the highest (known) energy input in the ISM in our galaxy. Any other acceleration process would therefore need an efficiency higher than 10%. On larger scales also AGN offer a high energy input in the ISM.

## 1.5 Galactic Sources of VHE γ-Rays

- The shock produced by supernova explosions can accelerate particles via diffusive shock acceleration (Fermi acceleration) (Bell 1978; Blandford & Eichler 1987). The predicted source spectrum of cosmic rays is a power law with index of about 2.1. Taking into account the energy dependent escape time of cosmic rays from the galaxy (power law with spectral index 0.6), the resulting power law spectral index fits nicely to the observed one of 2.7, see section 1.3.

If the SNRs are indeed the accelerators of cosmic rays, then they should also emit γ-rays through $\pi^0$ production in the collisions of the cosmic rays with the ambient matter, see e.g. Aharonian, Drury & Voelk (1994); Berezhko & Völk (1997, 2000); Malkov & Drury (2001). Assuming a spectral index of the cosmic ray power law of 2.1, Aharonian, Drury & Voelk (1994) predict the following integral γ-ray flux from an SNR (see also section 1.2.1):

$$\frac{dN_\gamma(E > E_0)}{dAdt} \approx 9 \cdot 10^{-11} \theta \left(\frac{E_0}{1\,\text{TeV}}\right)^{-1.1} \left(\frac{E_{\text{SN}}}{10^{51}\text{erg}}\right) \left(\frac{d}{1\,\text{kpc}}\right)^{-2} \left(\frac{n}{1\,\text{cm}^{-3}}\right) \text{cm}^{-2}\,\text{s}^{-1}\,, \tag{1.33}$$

where $\theta$ is the conversion efficiency of the total SNR power to cosmic rays, $E_{\text{SN}}$ is the total SN power, $d$ is the SN distance from the earth and $n$ is the (number) density of the ambient matter.

Therefore, for an SNR in the Sedov-Taylor phase at a distance of 1 kpc in an environment of density 1 cm$^{-3}$ (and assuming a cosmic ray acceleration efficiency of 10%) a VHE γ-ray flux above 1 TeV of about $9 \cdot 10^{-12}$ cm$^{-2}$s$^{-1}$ is expected, which is more than half of the flux of the Crab Nebula. Assuming an SNR stays for 10000 years in the Sedov phase and an SNR rate of 3 SNR/100 years there should be about 20 SNRs at a maximum distance of 4 kpc (with respect to the radius of the galaxy of 15 kpc) with minimum integral fluxes above 1 TeV of 3% of the flux of the Crab Nebula. To prove that SNRs are indeed the accelerators of the cosmic rays one has to observe enough SNRs in VHE γ-rays and show that the observed γ-radiation is due to hadronic interactions (see section 1.2).

The observations of hard X-rays are strong evidences for the existence of 100 TeV electrons in shell-type SNRs, see e.g. Koyama et al. (1995). The cut-off in the synchrotron spectrum is directly related to the maximum energy of the accelerated electrons. Electrons and nuclei are expected to be accelerated in a similar manner (Ellison & Reynolds 1991). At very high energies shell-type SNRs are detected with very high statistical significances such as the SNRs RXJ1713.7-3946 and RXJ0852.0-4622 (Vela Junior) by the CANGAROO (Enomoto et al. 2002; Katagiri et al. 2005; Enomoto et al. 2006) and HESS collaborations (Aharonian et al 2004c; Aharonian et al. 2005c). These observations prove that charged particles are accelerated in SNR shock fronts. Nevertheless, the nature of these VHE particles (electrons or hadrons) cannot be unambiguously identified yet. Therefore, deeper observations of the SNRs are necessary, especially the extension of the measured spectra from few GeV to several tens of TeV.

### 1.5.3 Pulsars and Pulsar Wind Nebulae

Pulsars are rapidly rotating neutrons stars which are produced in a type II supernova explosion. A pulsar wind nebula, PWN, (also known as a "plerion", Greek for "full") is a nebula emitting synchrotron radiation, which is powered by the relativistic wind of an energetic pulsar. Young pulsar wind nebulae are often found inside the shells of supernova remnants. The prototype pulsar wind nebula is the Crab Nebula.

The rotating strong magnetic field of the neutron star produces strong and variable electric fields (Goldreich & Julian 1969), where particles are accelerated to high energies. Due to the variable electric field, these particles (electrons and positrons) emit pulsed synchrotron radiation, see e.g. Harding (1981). Depending on the location of the acceleration region, so-called "polar cap" and "outer gap" models are distinguished. They predict $\gamma$-radiation with cut-offs at a few GeV or a few tens of GeV, respectively.

Almost all pulsars have rotational periods that are steadily increasing with time. This "spin-down" corresponds to a loss of rotational kinetic energy in the range up to $10^{39}$ erg/s. Most of this energy loss is thought to be dissipated by a magnetized wind of relativistic electrons and positrons, see e.g. Gaensler et al. (2000). At some distance from the pulsar, the pressure of the wind is eventually balanced by the external pressure, resulting in a strong stationary shock front, where particles (predominantly electrons and positrons) are accelerated to very high energies (in the Crab Nebula up to $10^{15-16}$ eV). These particles then emit synchrotron radiation and produce inverse Compton emission (see e.g. de Jager & Harding (1992); Harding (1996) and Aharonian, Atoyan & Kifune (1997)).

Pulsar wind nebulae are generally characterized by the following properties:

1. An increase in brightness towards the center, lack of a shell-like structure

2. A flux of highly polarized photons, the spectral index steepens from radio to X-ray energies due to synchrotron radiation losses

3. A steepening of the spectral index with distance from the pulsar due to smaller synchrotron lifetimes of the higher energy electrons (Safi-Harb 2004).

In 1989, steady VHE $\gamma$-ray emission ($E_\gamma \geq 1$ TeV) from the Crab Nebula was detected by the Whipple collaboration (Weekes et al. 1989), the first VHE $\gamma$-ray source. The Crab Nebula is the brightest stable VHE $\gamma$-ray source and remains the best-studied object of this class. It is commonly used for intercalibration between different $\gamma$-ray instruments.

### 1.5.4 $\gamma$-Ray Binaries

A recently discovered class of galactic VHE $\gamma$-ray sources are $\gamma$-ray binaries. Until now there are three known $\gamma$-ray binaries: PSR B1259-63 (Aharonian et al. 2005d), LS 5039 (Aharonian et al. 2005e) and LSI +61 303 (Albert et al. 2006d). These three binary systems consist of a stellar mass compact object (either a neutron star or a black hole of up to a few solar masses) and a massive star (10 to 20 solar masses) in an eccentric orbit around each other. According to Mirabel (2006) there are two models for the VHE $\gamma$-ray emission (for an illustration see figure 1.6):

- the microquasar jet model
- the pulsar wind model.

## 1.5 Galactic Sources of VHE γ-Rays

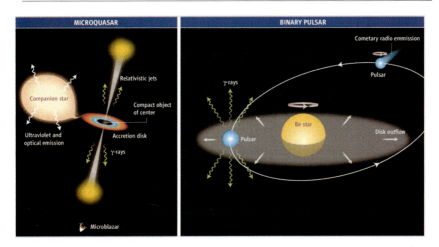

Figure 1.6: *Alternative models for VHE γ-ray binary systems: (Left) Microquasars are powered by the mass accretion from a large companion star onto a compact object (neutron star or stellar mass black hole). The accretion produces collimated jets like in the case of AGNs. γ-rays are produced in the jets. (Right) Pulsar winds are powered by the rotation of the neutron stars; the wind flows away to large distances in a comet-shaped tail. Interaction of this wind with the companion star outflow may produce the VHE γ-rays. Figure from Mirabel (2006).*

The left part of figure 1.6 shows the microquasar jet model: A normal star and either a black hole or a neutron star orbit around each other in a binary system. The companion star loses matter into an accretion disk around the compact object which is heated up to about $10^7$ K, see e.g. Mirabel & Rodriguez (1999) for a review. Part of the infalling matter emerges again in form of two relativistic jets, which emit radio waves and X-rays. In shock fronts within the jets charged particles may be accelerated to high energies (see section 1.4), which subsequently produce γ-rays via the inverse Compton effect or the interaction of hadrons, see section 1.2 and Paredes (2005).

Microquasars are named after the similarities with active galactic nuclei (AGNs), see section 1.6. Microquasars show the same three ingredients that make up radio-loud AGNs: a compact object, an accretion disc and relativistic jets (Mirabel et al. 1992; Mirabel & Rodriguez 1994). Hence, microquasars are galactic, scaled-down versions of AGNs, which have a stellar mass black hole instead of a super-massive black hole. The relative close distance of microquasars compared to AGNs make these objects ideal laboratories for the study of the physical processes of accretion and jets which determine the internal working of microquasars and AGNs. In some models (Heinz & Sunyaev 2002) microquasars could measurably contribute to the density of galactic cosmic rays.

The right part of figure 1.6 illustrates the pulsar wind model (see also section 1.5.3): The compact object is a young neutron star with a high spin down energy loss, which is transferred to a pulsar wind of non-thermal relativistic particles. When the pulsar wind interacts with the stellar wind of the companion star, strong shocks are created and charged particles can be accelerated to very high energies (see section 1.4). Again, $\gamma$-rays are produced via the inverse Compton effect or in the interaction of hadrons, see section 1.2.

### 1.5.5 Giant Molecular Clouds

The interstellar medium (ISM) consists of an extremely dilute plasma, gas and dust, which consists of a mixture of ions, atoms, molecules, larger dust grains, electromagnetic radiation, cosmic rays and magnetic fields. The matter consists, by mass, of about 99% gas and 1% dust (McKee & Ostriker 1977; Dyson 1997). The gas is roughly 90% hydrogen and 10% helium by number, with heavier elements present in trace amounts. The gas density varies between $10^{-4}\,\mathrm{cm}^{-3}$ in coronal gas and $10^5\,\mathrm{cm}^{-3}$ in dense molecular clouds. The hydrogen occurs in three different forms: as ionized plasma (H II regions), as atoms (H I regions), and as molecular clouds, which account for about half of the total hydrogen.

H II regions can readily be observed by their hydrogen spectral line emission (e.g. in the visible light). In an H II region, star formation is taking place. Young, hot, blue stars which have formed from the gas emit copious amounts of ultraviolet light, ionizing the hydrogen atoms around them. The radiation pressure from the hot young stars causes the H II regions to disperse at time scales of around a million years.

H I regions are colder and more dilute compared to H II regions. They do not emit visible light, but they are well detectable by the hyperfine structure emission line of 21 cm radio waves. Mapping H I emissions with a radio telescope is a technique used for determining the structure of spiral galaxies.

Giant molecular clouds (GMC) are very cold (a few tens of K) and dense clouds consisting mostly of molecular hydrogen. Collapses of parts of a GMC may lead to the birth of stars which subsequently heat the cloud to become an H II region. Molecular hydrogen is very difficult to detect as it does not emit any prominent lines in the electromagnetic spectrum (it has only a weak signature from the electric quadropol moment). Therefore, indirect methods using the detection of tracer molecules like CO are often applied, see e.g. Dame et al. (2001); Jackson et al. (2006).

CO is a polar molecule with a strong electric dipole rotational emission in the mm waveband, and is considered a reliable tracer of molecular hydrogen. The $J = 1 \leftrightarrow 0$ rotational transition of CO at about 115 GHz can be well observed. The emission is Doppler shifted according to the rotation speed of the GMC around the Galactic Center. Assuming a certain galactic rotation curve (see e.g. Clemens (1985)) one can estimate the distance of the GMC from the Galactic Center by the Doppler shift of the CO emission line. Molecular gas masses can be derived assuming a constant ratio between the molecular hydrogen column density $N(H_2)$ and the velocity-integrated (distance integrated) CO intensity (Dame et al. 2001):

$$N(H_2)/W_{\mathrm{CO}} = 1.8 \times 10^{20}\,\mathrm{cm}^{-2}\mathrm{K}^{-1}\mathrm{km}^{-1}\mathrm{s}^{-1} \ . \tag{1.34}$$

## 1.5 Galactic Sources of VHE γ-Rays

According to equation 1.33 (see also section 1.2.1) the expected γ-ray flux from a source depends on the total cosmic ray power at the source as well as on the density of target material. In the absence of a cosmic ray accelerator in the molecular cloud, the ambient sea of cosmic rays might produce detectable γ-ray signals only in the most dense and heavy molecular clouds. Contrary to that, a GMC near a cosmic ray accelerator could readily be a source of VHE γ-rays (Aharonian et al. 2001; Torres et al. 2003). One such possible GMC detected in VHE γ-rays is the region of the galactic disk close to the Galactic Center ($|b| < 0.2°$ and $|l| < 1.5°$), for which a correlation between the γ-ray brightness and target density is observed (Aharonian et al. 2006c).

### 1.5.6 The Galactic Center

The Galactic Center (GC) region contains many unusual objects which may be responsible for high-energy processes generating γ-rays (Aharonian & Neronov 2005; Atoyan & Dermer 2004; Horns 2004). The GC is rich in massive stellar clusters with up to 100 OB stars (Morris & Serabyn 1996), immersed in a dense gas. There are young shell-type supernova remnants e.g. G0.570-0.018 or Sgr A East, the pulsar wind nebula candidate G359.95-0.04 (Wang et al. 2006) and nonthermal radio arcs. The dynamical center of the Milky Way is associated with the compact radio source Sgr A*, which is believed to be a super massive black hole of about $3-4 \cdot 10^6 \, M_\odot$ (Ghez et al. 2000; Schödel et al. 2002; Ghez et al. 2003a,b; Eisenhauer 2005). Bower et al. (2004) have resolved the Galactic Center source at 7 mm wavelength, which yielded a radius of just 24 Schwarzschild radii (or about 2 AU). Within a radius of 300 pc around the Galactic Center a mass of about $3 \cdot 10^7 M_\odot$ is observed. An overview of the radio sources in the GC region is given in figure 1.7, while figure 1.8 shows a schematic diagram of the principal constituents of the Sgr A radio source at a much smaller scale. Some data about the GC are summarized in Table 1.1.

| | |
|---|---|
| (RA, Dec) of SgrA*, epoch J2000.0 | ($17^h 45^m 39.95^s$, $-29°00'28.2"$) |
| heliocentric distance | $8 \pm 0.4$ kpc (Reid 1993; Eisenhauer 2003, 2005) |
| mass of the black hole | $3.6 \pm 0.4 \cdot 10^6 M_\odot$ (Eisenhauer 2005) |

Table 1.1: Properties of the Galactic Center.

The γ-ray satellite EGRET has detected a strong source in direction of the GC, 3 EG J1745-2852 (Mayer-Hasselwander et al. 1998), which has a broken power law spectrum extending up to at least 10 GeV, with a spectral index of 1.3 below the break at a few GeV. Assuming a distance of 8 kpc from the GC, the γ-ray luminosity of this source is very large, i.e. $2.2 \cdot 10^{37}$ erg/s, which is equivalent to about 10 times the γ-ray flux from the Crab Nebula. However, an independent analysis of the EGRET data (Hooper & Dingus 2005) indicates a point source whose position is different from the GC at a confidence level beyond 99.9 %.

At energies above 200 GeV, the GC has been observed by the CANGAROO (Tsuchiya et al. 2004), VERITAS (Kosack et al. 2004) and HESS (Aharonian et al. 2004b) collaborations. The γ-ray spectra as measured by these experiments are displayed in figure 1.9 while figure 1.10 shows the different reconstructed positions of the GC source. Figure 1.11 shows

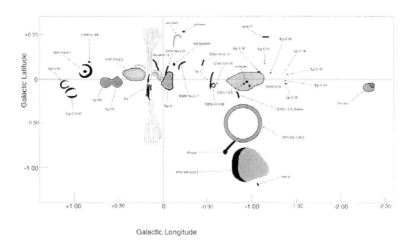

Figure 1.7: Overview about the radio sources near the Galactic Center (Lazio et al. 2005).

the total spectrum of the electromagnetic radiation from the Galactic Center (Aharonian & Neronov 2005). Recently, two more TeV sources and some diffuse $\gamma$-ray emission along the Galactic Plane have been reported by Aharonian et al. (2005b, 2006a,c).

The discrepancies between the measured $\gamma$-ray spectra could indicate inter-calibration problems between the IACTs. However, it could also indicate an apparent source variability at a timescale of about one year or it could be due to the different regions in which the signal is integrated.

In the GC region VHE $\gamma$-rays can be produced in different sources:

- the compact radio source Sgr A* (Aharonian & Neronov 2005)
- a possible AGN-like relativistic jet originating from the spinning GC black hole (Falcke et al. 1993; Falcke & Markoff 2000)
- the young SNR Sgr A East (Fatuzzo & Melia 2003; Yusef-Zadeh et al. 1999)
- the pulsar wind nebula candidate G359.95-0.04 (Wang et al. 2006; Hinton & Aharonian 2006)
- interaction between cosmic rays and the dense ambient gas within the innermost 10 pc region (Aharonian et al. 2006c)
- the non-thermal radio filaments (Pohl 1997)
- the central part of the Dark Matter halo (Prada et al. 2004).

## 1.5 Galactic Sources of VHE γ-Rays

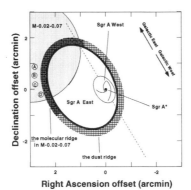

Figure 1.8: Schematic diagram showing the sky locations and rough sizes and shapes of the principle constituents of the Sgr A radio complex: The coordinate offsets are with respect to the compact nonthermal radio source Sgr A*, which coincides with the dynamical center of the galaxy, the super massive black hole candidate. Sgr A* is located at the center of the thermal radio source Sgr A West, which consists of a spiral-shaped group of thermal gas filaments. Sgr A West is surrounded by the molecular ring (also known as the circumnuclear disk), the radius of which is about 30". The nonthermal shell-like radio source Sgr A East is surrounding Sgr A West, but its center is offset by about 50". The nonthermal shell is surrounded by the dust and the molecular ridge. The molecular cloud M −0.02 −0.07 is located to the east of Sgr A East. At the eastern edge of the Sgr A East shell, a chain of H II regions (A-D) is seen. One arcminute corresponds to about 2.3 pc at the distance of 8 kpc. Figure from Maeda et al. (2002)

It is quite possible that some of these potential γ-ray production sites contribute comparably to the observed VHE γ-ray flux. For example, the young SNR Sgr A East is only located about 7 pc (about 0.05 deg) away from the Galactic Center (Yusef-Zadeh et al. 1999).

Production of VHE γ-rays within 10 Schwarzschild radii of a black hole (of any mass) could be copious because of effective acceleration of particles by the rotation-induced electric fields close to the event horizon or by strong shocks in the inner parts of the accretion disk. However, these energetic γ-rays generally cannot escape the source because of severe absorption due to interactions with the dense, low-frequency radiation through photon-photon pair production. Fortunately, the supermassive black hole in our Galaxy is an exception because of its unusually low bolometric luminosity. The propagation effects related to the possible cascading in the photon field may extend the high-energy limit to 10 TeV or even beyond (Aharonian & Neronov 2005).

Many proposed acceleration mechanisms of VHE γ-radiation in the Galactic Center are based on so-called advection dominated accretion flow (ADAF) models (Atoyan & Dermer 2004). A viable site of acceleration of highly energetic electrons could be the compact region

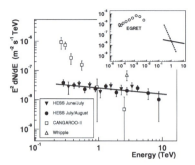

Figure 1.9: The VHE γ-ray flux from the Galactic Center as observed by the CANGAROO, VERITAS and HESS collaborations and by the EGRET experiment (Aharonian et al. 2004b). A clear discrepancy between the flux level and the spectral indices of the HESS and CANGAROO measurements are apparent.

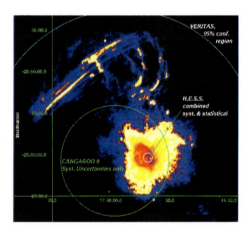

Figure 1.10: The VHE γ-ray source locations as measured by the IACTs Whipple, CANGAROO and HESS overlayed on a 90 cm radio map (Horns 2004).

within a few Schwarzschild radii of the black hole. In this case the electrons produce not only curvature radiation, which peaks around 1 GeV, but also inverse Compton γ-rays (produced in the Klein-Nishina regime) with the peak emission around 100 TeV. As these high-energy γ-rays cannot escape the source, the observed γ-rays would be due to an electromagnetic cascade.

## 1.5 Galactic Sources of VHE γ-Rays

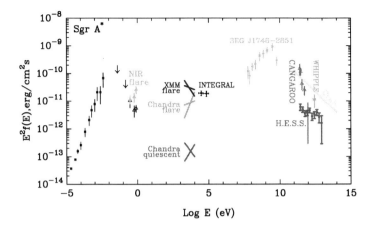

Figure 1.11: Total spectrum of the electromagnetic radiation from the Galactic Center, compiled by Aharonian & Neronov (2005).

Another scenario is related to accelerated protons, which produce γ-rays via the production and subsequent decay of neutral mesons like $\pi^0$s. Protons can be accelerated by the electric field close to the Schwarzschild radius of the black hole or by strong shocks in the accretion disk (Aharonian & Neronov 2005) to energies of about $10^{13}$ eV. In this case the γ-ray production is dominated by interactions of $10^{13}$ eV protons with the accretion plasma. This scenario predicts a neutrino flux which should be observable with northern neutrino telescopes like NEMO and Antares. It also predicts strong TeV–X-ray–IR correlations, for a deeper discussion see Aharonian & Neronov (2005).

If the protons are accelerated to energies as high as $10^{18}$ eV, detectable fluxes of $10^{18}$ eV neutrons are predicted (Aharonian & Neronov 2005). Neutrons of this energy travel a distance of about 8 kpc during their half-life time. A hint of an excess of highest energy neutrons from the GC has been reported by Hayashida et al. (1999).

### 1.5.7 The Galaxy in the Light of VHE γ-Rays

The accelerators of the cosmic rays can be searched for using their emitted electromagnetic radiation (and, in future, using neutrinos). According to section 1.2, the accelerated VHE electrons (and positrons) emit synchrotron radiation in the ambient magnetic fields (a continuum of radiation in the waveband between radio and hard X-rays), and VHE γ-rays via the IC upscattering of ambient photons. VHE protons produce, in their interactions with ambient gas, γ-rays with a peak intensity at about 70 MeV (half the mass of the $\pi^0$) and only a very weak synchrotron signal from secondary electrons. The acceleration sites of hadrons and electrons can be distinguished by their different overall multiwavelength emission, especially by the presence or absence of synchrotron emission. In case of a combined electron/proton acceleration, this source would emit inverse Compton VHE γ-rays and synchrotron radio to X-rays. Therefore, only the shape of the VHE γ-ray spectrum may allow to prove the acceleration of hadrons.

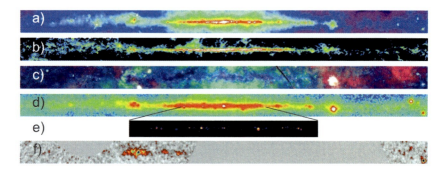

Figure 1.12: *Multiwavelength observations of the Milky Way (the Galactic Center is always in the middle of the images): a) Radio continuum (408 MHz) surveys with the Jodrell Bank, Bonn, and Parkes radio telescopes (Haslam et al. 1982). b) $^{12}CO\ J = 1-0$ spectral line emission (115 GHz) indicating the molecular hydrogen column density (Dame et al. 2001). c) Composite X-ray intensity observed by the Röntgen Satellite ROSAT, images in three soft X-ray bands centered at 0.25, 0.75, and 1.5 keV encoded in red, green, and blue color ranges, respectively (Snowden et al. 1997). d) Intensity of high-energy γ-ray emission above 300 MeV observed by the EGRET instrument on the Compton Gamma-Ray Observatory (Hartman et al. 1999). e) γ-ray intensity above 200 GeV from the HESS survey of the inner galaxy (Aharonian et al. 2006a). f) γ-ray emission significance map ($E_\gamma > 10$ TeV) using the Milagro detector (Goodman et al. 2006) (the GC and parts of the inner galaxy cannot be accessed by Milagro due to its location in the northern hemisphere).*

Unbiased surveys of the Galactic Plane in continuum radio and in X-rays have been available for some time. Two examples of such surveys are shown in figure 1.12a) and c). However, to find the exact sites of cosmic hadron acceleration, also unbiased surveys in the γ-ray band are necessary: In high energy γ-rays above 100 MeV the Galactic Plane was surveyed by the EGRET instrument (Hartman et al. 1999). Still, the majority of the galactic EGRET sources summarized in the 3rd EGRET catalogue (Hartman et al. 1999) remain unidentified, mainly due to the relatively poor spatial resolution of the instrument of about 1°. Only pulsars could by unambiguously identified by their pulsation period. There are some proposed associations of EGRET sources with SNRs and microquasars. Figure 1.12d) shows the map of the Galactic Plane in γ-rays above 300 MeV observed by EGRET. It shows a high level of diffuse emission along the Galactic Plane, which is assumed to be produced in interactions of cosmic rays with ambient gas and photon fields.

For γ-ray energies above a few TeV the full northern sky was surveyed by the Milagro water-Cherenkov detector (Atkins et al. 2004) and the Tibet air-shower array (Amenomori

## 1.5 Galactic Sources of VHE γ-Rays

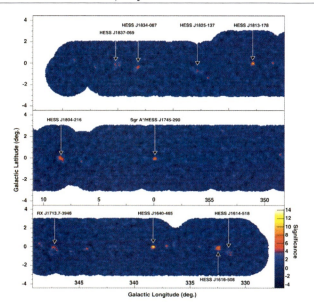

Figure 1.13: *Significance map for VHE γ-radiation from the inner galaxy (Aharonian et al. 2005a).*

et al. 2002). These experiments have a very large field of view (about 1 sr) but only a limited sensitivity and angular resolution. Figure 1.12f) shows a map of the Galactic Plane in terms of significance for the γ-ray emission above 10 TeV. In addition to the Crab Nebula the Milagro collaboration has reported evidence for γ-ray emission above 10 TeV along the Galactic Plane and from a source in the Cygnus region (Atkins et al. 2005; Fleysher et al. 2005; Goodman et al. 2006).

A first survey of the Galactic Plane with an IACT was carried out with the HEGRA (High Energy Gamma Ray Astronomy) telescopes array (Aharonian et al. 2002) in the range of galactic longitudes $-2° < l < 80°$. No sources were found at flux levels of 15% (outer galaxy) to 30% (inner galaxy) of the Crab Nebula flux. Recently, a scan of the inner galaxy ($|l| < 30°$) has been performed by the HESS collaboration (Aharonian et al. 2005a, 2006a). The scan had a sensitivity of 3% of the Crab Nebula flux above 200 GeV. Eight new sources were discovered in the original scan data with significances above 6 standard deviations, taking all trial factors of the different searched sky positions into account. Further observations increased the number of new significant sources to 15 (Aharonian et al. 2006a). Figure 1.12e) shows an expanded view of the VHE γ-ray intensity map of the inner part of the galaxy using the HESS telescope and figure 1.13 shows the corresponding significance map for VHE γ-ray emission.

In order to identify the nature of the newly discovered sources of VHE γ-rays they have to be firmly related to counter-parts observed in other wavelengths. Therefore, three conditions have to be fulfilled: (1) positional agreement between the sources (in the case of source extension also a morphological match) (2) the positional counter-part source should have a viable γ-ray emission mechanism (3) there should be a consistent picture of the multiwavelength emission of the source.

To search for possible accelerators of cosmic rays those γ-ray sources are most interesting, which are not associated with a pulsar (pulsar wind nebulae) and which do not have strong X-ray emission, which would indicate a dominantly leptonic origin of the γ-ray emission. From the eight significant sources there were two sources without any published counter-part candidate (Aharonian et al. 2005a): HESS J 1614-518 and HESS J 1813-178. Moreover, the source HESS J 1834-087 is positionally consistent with the SNR G 23.3-0.3, but without detected X-ray emission. This indicates only a low number of accelerated electrons and may hint to the acceleration of hadrons.

## 1.6 Extra Galactic Sources of VHE γ-Rays

In addition to the VHE γ-ray sources located in our galaxy, extragalactic astronomical objects can also be identified as sources of VHE γ-rays. Up to now all observed extra galactic sources of VHE γ-rays are Active Galactic Nuclei (AGN). However, there are theoretical models which predict also observable signals from γ-ray bursts (GRBs), starburst galaxies, clusters of galaxies and further, more exotic sources.

Taking their distance into account, AGNs are the most luminous sources confirmed of emitting γ-radiation. A unified model of AGNs was proposed by Urry & Padovani (1995) based on existing black hole models (Rees 1984): AGN are a class of galaxies which have supermassive black holes of $10^6$ to $10^9$ solar masses in their center. By the infall of matter onto the black hole a thin layer of hot plasma (the accretion disk) is formed. It approaches the black hole on spiral trajectories. The plasma emits a thermal spectrum peaking at X-rays. Often a pair of strongly collimated relativistic matter outflows ("jets"), perpendicular to the accretion disc, are observed. Most likely in shock fronts in these jets particles are accelerated to very high energies which subsequently emit radiation in the radio to within γ-ray waveband. Depending on the observation angle with respect to the jet axis, a rich phenomenology of AGNs can be observed leading to many classes and sub-classes of AGNs. For γ-ray astronomy, the most important ones are those where a jet points directly in the direction of the Earth, the so-called blazars. The first blazar (and the first extragalactic source ever) observed in VHE γ-ray was Mrk-421 (Punch et al. 1992). Recently, the radio-loud AGN M87 was detected in VHE γ-rays (Aharonian et al. 2003). It is the first AGN seen in VHE γ-rays with a jet off-axis.

Blazars show highly variable fluxes from the radio to the TeV regime. The emission level can vary by orders of magnitude in short time scales from minutes to months. Studying the γ-ray flux variations with time in different energy bands allows to set limits to an energy dependence of the speed of light due to possible quantum gravity effects, see e.g. Biller et al. (1999) and references therein.

Measurements of the γ-radiation from AGNs allow the study of matter accretion onto black holes and the jet phenomena. It is an interesting question whether the observed $\gamma$-radiation is produced by primary accelerated electrons or protons (see section 1.2). In the case of proton acceleration, AGNs could produce a substantial part of the observed Cosmic Rays on Earth. In the case of accelerated electrons the same electrons would emit synchrotron radiation in the ambient magnetic fields as well as scatter ambient photons up to very high energies. This scenario predicts strong correlations between the X-ray and VHE $\gamma$-ray fluxes (Fossati et al. 1998).

Another very interesting aspect of searching for far away extragalactic sources is the attenuation of the $\gamma$-rays on their way. Interstellar gas and dust can hardly stop VHE $\gamma$-rays; they easily penetrate a few g/cm$^2$ of matter, much more than what is found in interstellar gas clouds. However, VHE $\gamma$-rays can interact with the extragalactic background light (EBL) and produce electron positron pairs (Gould & Schreder 1966; Jelley 1966; Stecker et al. 1992). The higher the energy of the $\gamma$-ray, the lower the energy of the EBL photon can be resulting in a higher absorption probability of the $\gamma$-ray. A given EBL density leads to a maximum distance beyond which a $\gamma$-ray source cannot be seen any more (Fazio & Stecker 1970; Blanch & Martinez 2005), the so-called "$\gamma$-ray horizon". On the other hand, the EBL density can be determined by measuring the $\gamma$-ray spectra of sources with known distances/redshifts (see e.g. Dwek & Krennrich (2005)).

The density of the EBL is of significant cosmological interest, since it represents the light emission of galaxies summed over the entire history of galaxy formation since the Big Bang, see e.g. Kashlinsky (2005). Direct measurements (for a review see Hauser & Dwek (2001)) on or near Earth are very difficult, because of overwhelming amounts of foreground light originating in the solar system or in our own Galaxy.

## 1.7 Searches for $\gamma$-Rays from Dark Matter Annihilation

The existence of Dark Matter is well established on scales from galaxies to the whole universe, nevertheless, its nature is still unknown. Most of it cannot even be made of any of the known matter particles (for a review see e.g. Yao et al. (2006)). The most studied hypothesis for the nature of the Dark Matter is the existence of Weakly Interacting Massive Particles (WIMPs) left over from the big bang, see e.g. Jungman et al. (1996). These models are mainly motivated by extensions of the standard model of particle physics. Supersymmetric (SUSY) extensions of the standard model predict the existence of a promising Dark Matter candidate, the neutralino $\chi$, see e.g. Ellis et al. (1984). In most models its mass is below a few TeV. Models involving extra dimensions are also discussed like Kaluza-Klein Dark Matter (Bertone et al. 2003; Bergstrom 2004).

Any WIMP candidate (SUSY or not) may be detected directly via elastic scattering off nuclei in a detector on Earth. There are several dedicated experiments already exploiting this detection technique, but they have not yet claimed any strong and solid detection (for a review see Gascon (2005)). Complementary, WIMPs and especially SUSY neutralinos might annihilate in high-density Dark Matter environments and may be detected by their annihilation products. In particular, annihilation channels producing $\gamma$-rays are interesting because these are not deflected by magnetic fields and preserve the information of the original annihilation region, i.e. they act as tracers of the Dark Matter density distribution.

The expected mass range of SUSY neutralinos lies between about 50 GeV and a few TeV. Thus the continuum $\gamma$-ray spectra from potential SUSY neutralino annihilation coincides well with the accessible $\gamma$-ray energy region using IACTs, especially the MAGIC telescope. In section 1.7.1 possible targets for the observation of $\gamma$-rays from particle Dark Matter annihilation are given and the expected fluxes from these sources are estimated and compared to the sensitivity of the MAGIC telescope in section 1.7.2. These considerations have been published by Bartko et al. (2005).

### 1.7.1 Possible Observation targets

Neutralino annihilation can generate continuum $\gamma$-ray emission via the process $\chi\chi \to q\bar{q}$, see section 1.2.5. The subsequent decay of $\pi^0$-mesons created in the resulting quark jets produces a continuum of $\gamma$-rays.

According to equation 1.24 the expected $\gamma$-ray flux from neutralino annihilation is proportional to the Dark Matter density squared. Therefore, high density Dark Matter regions, which are possibly relatively nearby like the center of the Milky Way, its closest satellites and the nearby galaxies M31 and M87, are the most suitable places for indirect Dark Matter searches. Simulations and measurements of stellar dynamics indicate that the highest Dark Matter densities are found in the central part of galaxies and Dark Matter dominated dwarf-spheroidal-satellite galaxies (with large mass-to-light ratio). Numerical simulations in a Cold Dark Matter framework predict a few universal dark matter (DM) halo profiles (for example the so-called NFW profile (Navarro, Frenk & White 1997) or the Moore profile (Moore et al. 1998)). All of them differ mainly at low radii (pc scale), where simulation resolutions are at the very limit.

#### 1.7.1.1 Galactic Center

The presence of a Dark Matter halo in the Milky Way Galaxy is well established by stellar dynamics (Klypin et al. 2002). In particular, stellar rotation curve data of the Milky Way can be fit with the universal DM profiles predicted by simulations (Lokas et al. 2005; Fornengo et al. 2004; Evans et al. 2004). In addition, the Dark Matter may be compressed due to the infall of baryons to the innermost region (Prada et al. 2004) of a galaxy creating a central spike of the Dark Matter density. This central Dark Matter spike would boost the expected $\gamma$-ray flux from neutralino annihilation in the center of the galaxy. Although this model of baryonic compression is based on observational data and is in good agreement with cosmological simulations of the condensation of baryons (Gnedin et al. 2004), the existence of such a Dark Matter spike in the center of the Milky Way strongly depends on the Black Hole history during galaxy formation.

For a comparison between the expected $\gamma$-ray fluxes from neutralino annihilation and the MAGIC sensitivity, both the uncompressed NFW DM halo model (Navarro, Frenk & White 1997), see also Fornengo et al. (2004), and the adiabatic contracted NFW profile (Prada et al. 2004) are considered.

## 1.7 Searches for $\gamma$-Rays from Dark Matter Annihilation

### 1.7.1.2 Draco Dwarf Spheroidal and Nearby Galaxies

The Milky Way is surrounded by a number of small and faint companion galaxies. These dwarf satellites are by far the most Dark Matter dominated known objects, with Mass-to-Light ratios up to 300 times higher than the one of the sun. Draco is the dwarf satellite, with the largest DM content. Recently, a dwarf galaxy was discovered in the Ursa Major constellation, which may have an even higher mass-to-luminosity ratio (Willman et al. 2005).

DM density profiles derived from stars in the Draco dwarf cannot differentiate between cusped or cored profiles in the innermost region, as data are not available at small radial distances. Moreover, observational data disfavors tidal disruption effects, which may affect dramatically the DM distribution in Draco. For the discussions presented here the recent cusped DM model which includes new Draco rotation data (Lokas et al. 2005) is adopted.

Moreover, NFW models are adopted for the nearby galaxy M31 (Evans et al. 2004) and the Virgo Cluster (McLaughlin 1999). These profiles do not take into account any enhancement effect, like adiabatic contraction or presence of DM substructures.

### 1.7.2 Expectations for MAGIC

The $\gamma$-ray flux from neutralino annihilation can be derived by combining the SUSY predictions with the models of the DM density profile for a specific object. The SUSY predictions are taken from a detailed scan of the parameter space assuming Minimal Supergravity (mSUGRA), a simple and widely studied scenario for supersymmetry breaking (for details see Prada et al. (2004)). For a given choice of mSUGRA parameters the values of $m_\chi$, $\langle \sigma v \rangle$ and $N_\gamma$ (see equation 1.24) are determined and the consistency with all observational constraints is checked.

Comparing the expected $\gamma$-ray flux from neutralino annihilation in the considered candidate sources with the MAGIC sensitivity (see section 3.6.1), expected exclusion limits can be derived. Figure 1.14 shows the expected exclusion limits for 20 hours of MAGIC observations in the mSUGRA plane $N_\gamma(E_\gamma > E_{\text{thresh}})\langle \sigma v \rangle$ vs. $m_\chi$ for the four most promising sources considered. The nominal energy threshold $E_{\text{thresh}}$ has been conservatively assumed to be 100 GeV. It has been taken into account that the energy threshold and the effective collection area of the telescope are functions of the zenith angle of the observation, see section 3.6.5.

The expected fluxes are rather low and depend strongly on the innermost density region of the DM halos considered. The detection of a DM $\gamma$-ray signal from the Galactic Center may be possible in case of a very high density DM halo, like the one predicted by adiabatic contraction processes. The $\gamma$-ray flux from the Galactic Center as measured by the MAGIC and HESS telescopes is far above the theoretical expectations and extends to energies above 10 TeV, see section 4.1 and Aharonian et al. (2004b). Thus only part of this flux may be due to the annihilation of SUSY-neutralino Dark Matter particles (Horns 2004). Nevertheless, other models like Kaluza-Klein Dark Matter are not ruled out. It is interesting to investigate and characterize the observed $\gamma$-radiation to constrain the nature of the emission. The analysis of the MAGIC data of the Galactic Center is presented in section 4.1.

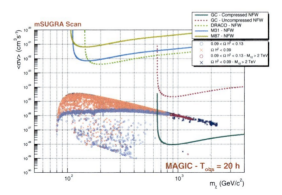

Figure 1.14: Expected exclusion limits for the four most promising sources of Dark Matter annihilation radiation for 20 hours of observation with MAGIC. The Galactic Center is expected to give the largest flux (lowest exclusion limits) amongst all sources. Figure from Flix (2005); Bartko et al. (2005).

In the long term Draco can be considered as a plausible candidate for Dark Matter inspired observations. Conservative scenarios give low fluxes which are not detectable by MAGIC in a reasonable observation time. However, there are several factors, like the clumping of Dark Matter, that might enhance the expected flux from neutralino annihilations in Draco. Other Dark Matter particles, like Kaluza-Klein particles, may produce higher $\gamma$-rays fluxes. Moreover, there are no known high energy $\gamma$-ray sources in the field of view (FOV) which could compete with the predicted $\gamma$-ray flux from Dark Matter annihilation.

## 1.8 Choice of Observation Targets for this Thesis

As shown above, the search for the sources of the cosmic rays is one of the main fundamental physics questions which can be answered using VHE $\gamma$-ray telescopes such as MAGIC. This question requires a profound understanding and modelling of the $\gamma$-ray emission mechanism of each of the galactic VHE $\gamma$-ray sources. The prototype sources of the hadronic cosmic rays would emit $\gamma$-rays with a hard spectrum of spectral index of around $-2.1$ reaching up to 100 TeV. Therefore, their flux may be below the EGRET sensitivity. Also, they may or may not emit in the radio and X-ray wavebands. An enhanced VHE $\gamma$-ray flux is predicted for sources near or in dense molecular clouds.

From the known galactic sources only supernova remnants and maybe the accreting supermassive black hole at the center of the galaxy release enough power for the acceleration of the galactic cosmic rays. Nevertheless, the cosmic rays may be accelerated in a still unknown population of sources.

## 1.8 Choice of Observation Targets for this Thesis

Therefore, I have proposed as Principal Investigator observations of three VHE $\gamma$-ray sources in our galaxy:

1. **Galactic Center**: The nature of the VHE $\gamma$-ray source at the Galactic Center is unknown due to uncertainties in the source localization and many possible source candidates. Also, the measured VHE $\gamma$-ray spectra by the different collaborations differ significantly which points to a source variability or instrumental problems. Conventional acceleration mechanisms for the VHE $\gamma$ radiation utilize the accretion onto the black hole, supernova remnants and PWN. The GC might be the brightest source of VHE $\gamma$-rays from particle Dark Matter annihilation. In this thesis the source location and the VHE $\gamma$-ray spectrum have been accurately determined. The integral VHE $\gamma$-ray flux of the source is shown to be stable in time.

2. **HESS J 1813-178**: When the source was discovered, no counter-parts in other wavelengths were known. This leads to speculations of a new class of dark particle (hadron) accelerators. Also no VHE $\gamma$-ray spectrum was known. Subsequent radio and X-ray observations showed spatially coincident sources, which may be due to an SNR. In this thesis the source location and the VHE $\gamma$-ray spectrum have been accurately determined, and the mutiwavelength emission of the SNR is modeled.

3. **HESS J 1834-087**: The source is spatially coincident with an SNR observed in radio, but not in X-rays. Initially, no VHE $\gamma$-ray spectrum was known. In this thesis the source location and morphology as well as the VHE $\gamma$-ray spectrum have been accurately determined. The VHE $\gamma$-ray source is shown to be spatially consistent with a dense molecular cloud.

There are observational constraints: Lying in the Galactic Plane, all three sources may only be observed at larger zenith angles up to 60° from La Palma. This leads to an increased energy threshold but also to an increased effective collection area of the MAGIC telescope. A dedicated analysis procedure for these observations was set up.

In addition, all three sources suffer from gradients in the sky brightness. In order to be least affected by these gradients and changing weather and telescope conditions, the sources are observed in a special observation mode, which also needed a dedicated analysis procedure.

# Chapter 2

# The MAGIC Telescope

The MAGIC (Major Atmospheric Gamma-ray Imaging Cherenkov) telescope is a new imaging Air Cherenkov telescope (IACT) on the Canary Island La Palma (28.8°N, 17.8°W, 2200 m above sea level). Its purpose is the ground-based detection of very high energy cosmic $\gamma$-radiation. The most important scientific objective is the understanding of the origin of the acceleration and reaction mechanisms of very high energetic particles in astronomical objects - the search for the accelerators of the cosmic rays.

The cosmic $\gamma$-rays cannot penetrate the earth's atmosphere and thus cannot be directly measured on the ground (see section 2.1.2). They interact with the nuclei of the atmosphere and can only be observed directly by expensive satellite-born telescopes. Due to their small dimensions satellites are presently only sensitive up to $\gamma$-ray energies of some tens of GeV from the strongest sources.

Nevertheless, VHE $\gamma$-rays, as well as hadronic cosmic rays, can be indirectly detected from the ground: When the cosmic $\gamma$-rays interact with nuclei of the earth's atmosphere, they produce so-called shower cascades consisting of thousands of electrons and positrons. These particle showers penetrate the earth's atmosphere typically by several kilometers, with the shower maximum typically 10 km above the sea level. Those electrons and positrons of the shower which are faster than the speed of light in air, emit Cherenkov light in the blue to UV spectral range, which is being observed by the MAGIC telescope: In the camera of the MAGIC telescope one obtains an encoded image of the shower (due to the Cherenkov angle). An analysis of this image allows to determine the arrival direction and the energy of the primary $\gamma$-ray. From the analysis of many showers a map of the $\gamma$-ray source can be obtained. Previous generation Cherenkov telescopes were able to observe $\gamma$-radiation only above 300 GeV. The aim of MAGIC is to cover the unexplored part of the electromagnetic spectrum between 30 and 300 GeV.

The general technique of Imaging Air Cherenkov telescopes, including the physical processes of air showers and their imaging by telescopes, is presented in section 2.1. Section 2.2 describes the hardware layout of the MAGIC telescope, while section 2.3 presents the modus of observation with the MAGIC telescope. Finally, section 2.4 describes the construction of a second MAGIC telescope in La Palma, intended for stereoscopic observations of air showers.

## 2.1 The Imaging Air Cherenkov Technique

This section will briefly summarize the general principles of imaging air Cherenkov telescopes and of the physical processes exploited by the technique. It is structured as follows: First, section 2.1.1 reviews the interactions of high energy particles within air. Thereafter, section 2.1.2 discusses the development of air shower cascades and section 2.1.3 describes the physical process of Cherenkov light emission. Finally, the imaging of the Cherenkov light from an air shower by a telescope is explained in section 2.1.4.

### 2.1.1 Interactions of High Energy Particles within Air

To describe the interaction of particles with matter, one considers interactions via the strong and the electromagnetic force. The weak force has to be considered only in the decays of the produced particles.

#### 2.1.1.1 Electromagnetic Interactions of Charged Particles within Air

There are five processes of electromagnetic interactions of charged particles within air:

- Bremsstrahlung
- Cherenkov radiation
- Ionization
- Excitation
- Photo production.

In the field of an "air" nucleus (typically from N, O, C, Ar), charged particles may radiate photons, which is known as **bremsstrahlung**. The energy loss of charged particles due to **bremsstrahlung** for high energies can be described by (see e.g. Fernow (1986); Grupen et al. (1996)):

$$-\frac{dE}{dx} = 4\alpha N_A \rho \frac{Z^2}{A} z^2 r_e^2 \left(\frac{m_e}{m}\right)^2 E \ln \frac{183}{Z^{1/3}}, \quad (2.1)$$

where $\alpha$ is the fine-structure constant, $N_A$ is the Avogadro constant, $\rho$, $A$ and $Z$ are the average density, atomic mass and charge of the absorber (air), $m_e$ is the electron mass, $r_e$ is the classical electron radius and $z$, $m$ and $E$ are the particle charge, mass and energy, respectively. Because of the small electron mass, bremsstrahlung plays a very important role for the energy loss of electrons. For electrons equation 2.1 is usually written as:

$$-\frac{dE}{dx} = \frac{E}{X_0} \quad \text{with} \quad X_0^{-1} = 4\alpha N_A \rho \frac{Z^2}{A} r_e^2 \ln \frac{183}{Z^{1/3}}, \quad (2.2)$$

where $X_0$ is the radiation length of the absorber. After having traveled a distance $X_0$ (in air: $X_0 \rho = 36.66 \frac{g}{cm^2}$, corresponding to $X_0 \approx 300$ m for standard pressure at sea level (Yao

## 2.1 The Imaging Air Cherenkov Technique

et al. 2006)) an electron has lost on average all but $1/e$ of its initial energy. For particles other than electrons bremsstrahlung plays a minor role up to energies of about 1 TeV.

The cross-section for **photo production** of hadrons is about three orders of magnitude lower than the one for bremsstrahlung (Eidelman et al. 2004). Thus photo production can be neglected. The energy loss due to **Ionization** and **Excitation** is described by the Bethe-Bloch formula, see e.g. Eidelman et al. (2004), but is not useful for a measurement. The **Cherenkov effect** on the other hand is essential for the signal production of the MAGIC telescope, see section 2.1.3.

### 2.1.1.2 Interactions of Photons within Air

For photon energies larger than a few MeV, **pair production** is the dominating interaction process. Here, the photon converts into an electron-positron pair in the electric field of an air nucleus. At lower photon energies the photon is likely to scatter off the quasi-free atomic electrons of the air, known as **Compton scattering**. By virtue of the **photo electric effect**, the photon is completely absorbed by an atomic electron, which is the main process for photon energies below a few keV.

The intensity $I$ of a photon beam as a function of the distance traveled within air obeys an exponential law:

$$I(x) = I_0 e^{-\mu \rho x} , \qquad (2.3)$$

with $I_0$ being the initial intensity, $\mu$ the mass absorption coefficient, $\rho$ the mass density and $x$ the distance traveled (thickness of the air). For photon energies above a few MeV the mass absorption coefficient can be approximated as (see e.g. Rossi (1965); Fernow (1986); Kleinknecht (1998)):

$$\mu \rho \simeq \frac{7}{9 X_0} , \qquad (2.4)$$

where $X_0$ is the radiation length defined above. Thus in a layer of air of one radiation length thickness a high energy photon converts into an electron positron pair with a probability of $1 - e^{-\frac{7}{9}} \simeq 54\%$. The thickness of the total atmosphere at sea level and normal pressure ($p = 1.013 \cdot 10^5$ Pa ) corresponds to about 28 $X_0$.

### 2.1.1.3 Strong Interactions of Hadrons within Air

There are many exclusive processes corresponding to specific hadronic final states, in which hadrons can interact with the nuclei of air. More than 300 such exclusive processes contribute only about 0.1% each to the total cross section of the interactions (Wigmans 1987). In principle all these processes are described by QCD, the theory of strong interactions, but up to now the theoretical modeling of these processes is still under development.

Above the nuclear resonances corresponding to initial state energies of about a GeV the cross section for hadron nucleus collisions shows only a weak energy dependence. The probability $p(x)$ that a high energy hadron travels a distance $x$ in the absorber *without* interacting with one of the nuclei is given by:

$$p(x) = \exp\left(-\frac{x}{\lambda_I}\right), \qquad (2.5)$$

where $\lambda_I$ is the nuclear interaction length (in air: $\lambda_I \rho = 90.0 \frac{g}{cm^2}$, or $\lambda_I \approx 740$ m for standard pressure at sea level (Yao et al. 2006)). The thickness of the total atmosphere at sea level at normal pressure ($p = 1.013 \cdot 10^5$ Pa, $\rho = 1.205 \cdot 10^{-3}$ g/cm$^3$) corresponds to about 11 $\lambda_I$.

### 2.1.2 Development of Air Shower Cascades

When high energy particles interact within air, secondary particles are produced which carry away part of the energy of the incoming particle. These secondary particles either decay or interact again with the air nuclei and produce tertiary particles and so on. Thus the primary energy is shared amongst a growing number of particles. This mechanism continues until the energy of the particles is not sufficient to produce new ones. The whole process from the first interaction of the incoming particle until the complete energy loss of the last particle is called a shower cascade.

According to their origin, one distinguishes between electromagnetic (initiated by electrons and photons) and hadronic (initiated by hadrons) showers.

#### 2.1.2.1 Electromagnetic Showers

Electromagnetic showers develop through the electromagnetic processes of bremsstrahlung and pair production. Initial high energy electrons (or positrons) radiate a bremsstrahlung photon in the electric field of a nucleus of the absorber material which then converts into a new electron positron pair and so on. The shower reaches its maximum particle multiplicity ("shower maximum"), when

- the energy of the electrons and positrons falls below the critical energy $E_c$ (81 MeV in air) such that they lose more energy by ionization than by bremsstrahlung and

- the photons have an energy below the threshold for pair production.

In figure 2.1 a schematic picture of an electromagnetic shower is shown. The size of electromagnetic showers is determined by the radiation length of the absorber material: About 98% of the energy of the incident particle is contained in a length of (Grupen et al. 1996):

$$L(98\%) = 2.5 X_0 \left[\ln\left(\frac{E}{E_c}\right) + C\right], \qquad (2.6)$$

where $E$ is the energy of the incoming particle. For photons, $C = +0.5$ and for electrons and positrons, $C = -0.5$. A shower of a 1 TeV photon in air with vertical incidence has an $L(98\%)$ of about 7.5 km. The transverse extension of an electromagnetic shower is

## 2.1 The Imaging Air Cherenkov Technique

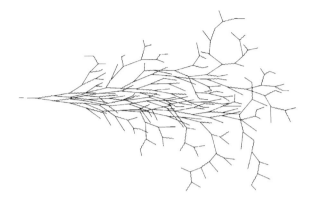

Figure 2.1: *Schematic picture of an electromagnetic cascade (Grupen et al. 1996). Straight lines represent electrons and positrons, while curly lines represent photons.*

often described in terms of the Molière-Radius $R_\text{M}$ (see e.g. Scott (1963); Eidelman et al. (2004)):

$$R_\text{M} = \frac{m_e c^2 \sqrt{\frac{4\pi}{\alpha}}}{E_c} X_0 \simeq 21.2 \text{MeV} \cdot \frac{X_0}{E_c} \, , \qquad (2.7)$$

About 95% of the energy of the shower is deposited in a cylinder of radius $2R_\text{M}$ ($R_\text{M} \simeq 75$m in air for standard pressure at sea level) around the axis of the shower (Grupen et al. 1996).

#### 2.1.2.2 Hadronic Showers

A hadronic shower is initiated by an incoming hadron which interacts with a nucleus of the absorber material via the strong force. In this reaction new hadrons are produced, which themselves scatter inelastically off atomic nuclei and thus produce further hadrons until the energy of the hadrons is below the production threshold for new hadrons. However, in addition to this purely hadronic process one also has to consider three other processes:

- Electromagnetic decays, especially of $\pi^0$s
- Weak decays of hadrons
- Ionization of the air.

In figure 2.2 a schematic picture of the development of a hadronic shower is shown.

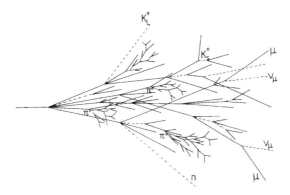

Figure 2.2: *Schematic picture of a hadronic cascade (Grupen et al. 1996). The full lines represent charged particles which may radiate Cherenkov light, whereas the broken lines represent neutral particles which do not contribute to the Cherenkov signal. Also electromagnetic sub-showers due to decays of neutral pions into two photons are shown.*

### 2.1.2.3 Differences between Electromagnetic and Hadronic Showers

For $\gamma$-ray astronomy with IACTs the differences between $\gamma$-ray initiated electromagnetic air showers and the background of hadronic air showers from charged cosmic rays are of major importance.

Due to a larger transverse momentum transfer in the hadronic interactions hadronic showers are broader and also more irregular due to the different interaction and decay processes, compared to electromagnetic showers. Figure 2.3 shows a $\gamma$- and a proton-initiated air shower for the same first interaction hight of 30 km above sea level. The $\gamma$-ray shower is more concentrated and regular compared to the proton shower. Half of the Cherenkov light emission from a $\gamma$-ray shower arises within 21 m of the axis, while a proton shower emits half of its light within 70 m of the axis (Hillas 1995).

Very important is also the production of $\pi^0$s in hadronic showers, which decay after a mean life time of about $10^{-16}$ seconds immediately into two photons which then initiate an electromagnetic sub-shower. As roughly 1/3 of the produced pions in each hadronic interaction are $\pi^0$s, a substantial amount of the initial energy of the incoming hadron is deposited in electromagnetic sub-showers. In some cases only these electromagnetic sub-showers are seen by the telescope; the rest of the produced particles either do not illuminate the telescope with Cherenkov light or do not emit any Cherenkov light at all. These electromagnetic sub-showers are an irreducible background to the electromagnetic showers from primary $\gamma$-rays.

## 2.1 The Imaging Air Cherenkov Technique

Figure 2.3: *γ-ray shower vs. proton shower: $E = 100$ GeV, dark color corresponds to high particle density, fixed first interaction height at 30 km (Schmidt 2005).*

### 2.1.3 The Emission of Cherenkov Light

A charged particle emits Cherenkov radiation if its velocity is greater than the local group velocity of light. The energy loss due to this process is negligible compared to the energy loss due to ionization, but it is the basis of the detection principle of air showers by IACTs, see e.g. Leo (1994).

The Cherenkov light is emitted in a cone with a half-angle $\theta_c$ for a particle with velocity $v = \beta c$ in a medium with refraction index $n$ (for air at standard pressure $(n-1) \times 10^6 = 293$):

$$\theta_c = \arccos\left(1/\beta n\right). \tag{2.8}$$

The threshold velocity $\beta_{\text{thr}} c$ for Cherenkov radiation is given by:

$$\beta_{\text{thr}} = \frac{1}{n}. \tag{2.9}$$

The number of photons produced per unit path length of a particle with charge $ze$ and per unit wavelength interval of the produced photons is (Eidelman et al. 2004):

$$\frac{d^2 N}{dx\, d\lambda} = \frac{2\pi \alpha z^2}{\lambda^2} \left(1 - \frac{1}{\beta^2 n^2(\lambda)}\right). \tag{2.10}$$

Due to the $1/\lambda^2$ dependence of the number of emitted photons and the $\lambda$-dependence of $n$, the peak of the $d^2N/(dx\, d\lambda)$ distribution lies in the UV region. Strong absorption processes of UV radiation in the atmosphere (mainly by ozone) lead to a maximum number of Cherenkov photons observed in the blue region of the electromagnetic spectrum. Moreover, Mie and Rayleigh scattering further reduce the number of Cherenkov photons reaching the telescope. Figure 2.4 shows the spectrum of Cherenkov light at the shower maximum (dashed curve) and after traveling down to 2 km altitude (full curve) (Barrio et al. 1998).

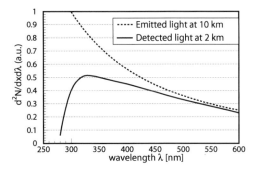

Figure 2.4: *Spectrum of Cherenkov light at the shower maximum (dashed curve) and after traveling down to 2 km altitude (full curve). Figure from Barrio et al. (1998).*

### 2.1.4 Imaging of the Cherenkov Light from an Air Shower by a Telescope

Figure 2.5 shows the working principle of an air Cherenkov telescope: A very high energy $\gamma$-ray entering the earth's atmosphere initiates a shower cascade of electrons and positrons with a shower maximum about 10 km above sea level (for an energy of 1 TeV). Electrons radiate Cherenkov light in a cone of about 1° half-angle which illuminates an area of around 120 m radius on the ground. Figure 2.6 presents simulated lateral distributions of the Cherenkov light density on a height of 2200 m above sea level from a 100 GeV $\gamma$-ray shower and a 400 GeV proton induced shower for vertical incidence (Barrio et al. 1998). In case the MAGIC telescope is inside the illuminated area, it collects part of the Cherenkov light with its mirrors and projects a shower image onto the PMT camera. The main background to the $\gamma$-ray showers originates from much more frequent showers induced by isotropic hadronic cosmic rays.

The Cherenkov photons arrive within a very short time interval of a few nanoseconds at the telescope camera (about 2 ns full width at half maximum (FWHM) for $\gamma$-ray showers, see section 5.1) . Using fast light sensors such as photomultiplier tubes (PMTs), one can trigger on the coincident light signals in different camera pixels. Exposure times (signal integration times) in the order of ten nanoseconds help to suppress the background from the light of the night sky (LONS). In general, IACTs can only be operated in dark conditions, i.e. moonless nights. The sensitivity of the PMTs of the MAGIC telescope has been adjusted such that it can also be operated during partial moon-shine (Barrio et al. 1998).

## 2.2 The Hardware Layout of the MAGIC Telescope

MAGIC (see e.g. Baixeras et al. (2004); Cortina et al. (2005) for a detailed description) is the largest single-dish Imaging Air Cherenkov Telescope (IACT) in operation. It is

## 2.2 The Hardware Layout of the MAGIC Telescope

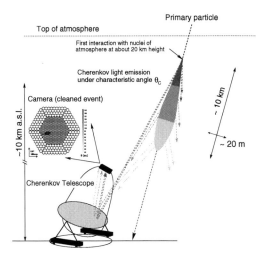

Figure 2.5: *IACT principle: A cosmic high energy $\gamma$-ray penetrates in the earth's atmosphere and initiates a shower cascade of electrons and positrons, which radiate Cherenkov light. This light is collected and focussed onto the camera, providing an image of the air shower. Picture adapted from Commichau (2004).*

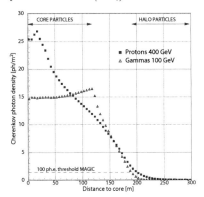

Figure 2.6: *Simulated lateral distribution of Cherenkov light density on a height of 2200 m above sea level from a 100 GeV $\gamma$-ray shower and a 400 GeV proton induced shower for vertical incidence (Barrio et al. 1998).*

located on the Canary Island La Palma (28.8°N, 17.8°W, 2200 m above sea level). There are four Cherenkov telescope observatories of the latest generation operating world-wide: CANGAROO (Australia, Kabuki et al. (2003)), HESS (Namibia, Aharonian et al. (2006d)), MAGIC (La Palma, Barrio et al. (1998)) and VERITAS (USA, Holder et al. (2006)).

The first design study of the MAGIC telescope was finished in 1998 (Barrio et al. 1998). In 2001 the production of the telescope components and the installation at La Palma started. The telescope frame was completed at the end of 2001 and the inauguration was celebrated in 2003. Thereafter, the telescope with all its sub-systems was thoroughly tested. In 2004 the first scientific observations were carried out (Albert et al. 2006c), and in spring 2005 the regular cycle 1 observations started.

Figure 2.7: *Picture of the MAGIC telescope: The tessellated mirror and the light-weight space frame construction can be seen.*

Figure 2.7 shows a picture of the MAGIC telescope. The mirror diameter is 17 m, corresponding to a total mirror area of 239 m$^2$. The camera of the MAGIC telescope with a mean field-of-view of 3.5° ($\sim$1 m camera diameter at 17 m focal length) consists of a densely packed matrix of 576 photomultiplier tubes (PMTs), which record the Cherenkov light. The PMTs have an enhanced quantum efficiency and 1.0 - 1.2 ns FWHM response to sub-ns input light pulses (Ostankov et al. 2000). The fast analog PMT signals are transported via 162 m long optical fibers to the trigger electronics and are read out by a 300 MSamples/s FADC system. For the images of the air showers exposure (signal integration) times of around 10 ns are used. The analysis of the form and orientation of the shower pictures permits to separate $\gamma$-ray initiated air showers efficiently from backgrounds (mainly showers of the charged cosmic rays).

This section is structured as follows: First, in section 2.2.1 the site location of the MAGIC telescope is described. Thereafter, details of important hardware parts of the telescope are presented: the telescope mechanics and the drive system (section 2.2.2), the mirror system (section 2.2.3), the MAGIC camera (section 2.2.4), the trigger system

## 2.2 The Hardware Layout of the MAGIC Telescope

(section 2.2.5), the data acquisition and signal processing system (section 2.2.6) and the calibration system (section 2.2.7). Finally, in section 2.2.8 an overview of the Monte-Carlo simulations of the MAGIC telescope is given, which is necessary to estimate the telescope response to $\gamma$-rays as the telescope response cannot be calibrated in test beams.

### 2.2.1 Site Location

The MAGIC telescope is located 2200 m above sea level at the Roque de los Muchachos Observatory on the Canary Island of La Palma (28.8°N, 17.8°W). This site was chosen for its height, its clear, cloudless nights and low humidity (Barrio et al. 1998). Moreover, the site has successfully hosted in the past the predecessor experiment HEGRA (see e.g. Daum et al. (1997)).

### 2.2.2 Telescope Mechanics / The Drive System

The MAGIC telescope has a 17 m diameter parabolic mirror dish with a focal length of 17 m, which is supported by a light-weight space frame of carbon fiber reinforced plastic tubes. This light-weight telescope design in an Altitude-Azimuth mount together with powerful servo-motors enables the MAGIC telescope to be repositioned to an arbitrary sky position in about 30 s. Figure 2.8 shows a technical drawing of the MAGIC telescope (Barrio et al. 1998). The telescope design was mainly done in the mechanical department of the MPI in Munich.

The tracking system of the MAGIC telescope is described in detail by Bretz et al. (2003). The pointing direction of the telescope is measured using shaft encoders with a resolution of 14 bit, corresponding to a telescope positioning accuracy of about 2'. Due to telescope structure deformations, however, there are sometimes larger deviations of the actual telescope pointing from the intended position, especially during the culmination of a tracked source. These positioning deviations can be monitored and corrected for using the MAGIC starfield monitor (Riegel et al. 2005): A video camera is attached to the center of the mirror dish of the MAGIC telescope. It compares monitor LEDs mounted on the PMT-camera frame and stars with well-known positions from the celestial background. From the position of the camera LEDs relative to the stars the actual pointing position of the MAGIC telescope can be determined with an accuracy of about $0.01° = 36''$.

### 2.2.3 The Mirror System

The MAGIC telescope has a 17 m diameter tessellated parabolic reflector (Bigongiari et al. 2004) to exploit the short duration of Cherenkov flashes. The mirror dish is segmented into 956 square tiles (50 cm×50 cm), corresponding to a total mirror area of 239 m$^2$. Each tile is an all aluminum spherical mirror with curvature radius ranging from 36.6 m, at the paraboloid rim, to 34.1 m at its center. The all aluminum mirrors are an innovation in $\gamma$-ray astronomy. They have a comparable optical quality as the conventional glass mirrors used in $\gamma$-ray astronomy while their weight is much less. The mirrors are equipped with an integrated printed-circuit board heating used for de-icing. The mirror system has a focal

Figure 2.8: *Technical drawing of the MAGIC telescope (Barrio et al. 1998).*

length of 17 m and therefore a focal length to diameter ratio $f/d = 1.0$. Four mirror facets are fixed on one panel.

In order to compensate zenith angle dependent deformations of the 17 m mirror dish the mirror panels can be readjusted (Garczarczyk et al. 2003). Each panel can be moved by two actuators. The positioning is monitored and adjusted using the laser spot on the camera lid from a laser pointer fixed to each panel. The optical point spread function (PSF) is the resulting image in the focal plane of a point light source at infinity produced by all mirrors. This image is well contained within the camera pixel size of 0.1 degrees. The optical PSF depends on the distance of the image from the camera center due to spherical aberrations of the reflector.

### 2.2.4 The MAGIC Camera

The camera of the MAGIC telescope serves to capture and record the Cherenkov images of air showers. Typical observations of the MAGIC telescope record images at a rate of around 300 Hz with negligible dead time. A schematics of the MAGIC camera layout is shown in figure 2.9. The camera of the MAGIC telescope (Cortina et al. 2003) has the following features:

- $\sim 1$ m diameter at 17 m focal distance, $\sim 3.5°$ field-of-view (FOV)

- inner hexagonal area of 397 hemispherical bialkali photocathode PMTs (ET 9116A) of 1 inch diameter, equivalent to 0.1° FOV each (including Winston cone light col-

## 2.2 The Hardware Layout of the MAGIC Telescope

lectors), surrounded by 179 PMTs (ET 9117A) of 1.5 inch diameter, equivalent to (including Winston cone light collectors) 0.2° FOV (Ostankov et al. 2000)

- PMTs are coated with a special lacquer which enhances the quantum efficiency (QE) up to 30% (Paneque et al. 2004)

- Dedicated light collectors (Winston cones, Winston (1970)) which reduce the dead area between the PMTs. These cones are specially designed to maximize the number of photon trajectories that cross the hemispherical photocathode twice to enhance further the QE. Moreover, these light collectors shield the PMTs from light from outside the telescope dish.

- Ultrafast and very low-noise transimpedance pixel pre-amplifier (bandwidth $\sim$1 GHz).

- The analog electrical PMT signals are transformed to analog optical signals by vertical cavity surface emitting lasers (VCSELs) and transported via 160 m long optical fibers to the trigger and read-out electronics in the counting house.

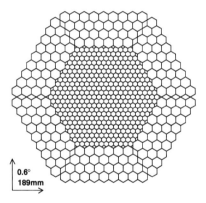

Figure 2.9: *Schematic picture of the MAGIC PMT camera. It consists of 397 inner pixels of 0.1° FOV each, surrounded by 179 outer pixels of 0.2° FOV. The total FOV is about 3.5°. The length of the arrows shows a size on the camera of 189 mm, equivalent to 0.6°.*

In general, a large FOV is desirable to observe extended sources, for sky scans and the serendipitous observations of unknown sources. A fine pixelization (the pixel size should not be bigger than the instrument PSF) is needed to accurately reconstruct the air shower parameters from the camera pictures. The chosen camera layout of 3.5° FOV with 577 pixels is a compromise between large FOV, fine pixelization and cost requirements.

## 2.2.5 The Trigger of MAGIC

The purpose of the trigger system is to recognize online air shower candidates by their coincident signals in adjacent pixels of the MAGIC camera and to initiate the read-out of the full camera information for air shower candidate events.

The trigger of the MAGIC telescope is a sophisticated three-level trigger system (levels 0, 1, 2) with programmable logic. At 0th level, for each channel a discriminator with an adjustable threshold indicates the presence of a significant signal above the noise level. The first level trigger applies tight time coincidences and a simple N-next-neighbor logic to the level zero outputs in any of 19 overlapping trigger cells, each comprising 36 pixels out of the 325 pixels of the trigger region (corresponding to a trigger FOV of about 0.9°, see the position of the 19 trigger cells in figure 2.10). In the case that N compact nearest neighboring pixels show a signal exceeding an adjustable threshold in a tight coincidence time interval of 5 ns, the level 1 trigger outputs a trigger signal.

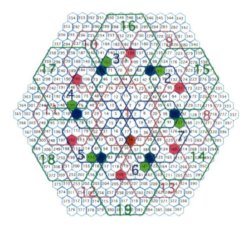

Figure 2.10: *Schematics of the trigger: There are 19 overlapping trigger cells of 36 pixels each, corresponding to a total of 325 pixels in the trigger. The location of these 325 trigger channels is indicated (only inner camera pixels are used for the trigger). There is a positive trigger level 1 decision if N compact nearest neighboring pixels in any of the 19 trigger cells exceed an adjustable threshold in a tight coincidence time interval of 5 ns.*

At trigger level 2 for each of the 325 pixels in the trigger region a threshold bit (set if the preset analog threshold is exceeded) is available. Using fast programmable electronics a pattern recognition algorithm is applied to this topological information and the most likely $\gamma$-ray candidate showers can be filtered out. Up to now, the second level trigger has not yet been used in the standard data taking procedure.

## 2.2 The Hardware Layout of the MAGIC Telescope

### 2.2.6 Data Acquisition and Signal Processing of the MAGIC Telescope

The MAGIC read-out chain, including the PMT camera, the analog-optical link, the majority trigger logic and FADCs, is schematically shown in figure 2.11. The response of the PMTs to sub-ns input light (the Cherenkov light pulse of a $\gamma$-ray shower has a FWHM of about 2 ns, see section 5.1) shows a pulse of FWHM of 1.0 - 1.2 ns and rise and fall times of 600 and 700 ps, respectively (Ostankov et al. 2000). By modulating vertical cavity surface emitting laser (VCSEL) diodes ($>$ 1 GHz bandwidth) in amplitude the ultra-fast analog signals from the PMTs are transferred via 162 m long, multi mode graded index 50/125 $\mu$m diameter optical fibers to the counting house (Lorenz et al. 2001). After transforming the light back to an electrical signal, the original PMT pulse has a FWHM of about 2.2 ns and rise and fall times of about 1 ns. Subsequently, the signal is split: One branch goes to a discriminator with a software adjustable threshold generating the signal for the trigger system. The signal in the second branch is sent to the 300 MSamples/s 8 bit FADC system (Goebel et al. 2003).

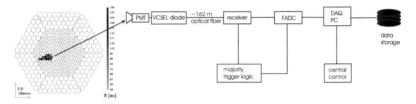

Figure 2.11: *Current MAGIC read-out scheme: the analog PMT signals are transferred via an analog optical link to the counting house where, after the trigger decision, the signals are digitized by a 300 MSamples/s FADCs system and written to the hard disk of a DAQ PC.*

In order to sample the fast pulse ($\sim$ 2.2 ns FWHM) with the present 300 MSamples/s FADC system, the pulse is stretched to a FWHM of $>$ 6 ns (the original pulse is folded with a stretching function of about 6 ns). This implies a longer integration of the light of the night sky (LONS) and thus the performance of the telescope on the analysis level is degraded (but see chapter 5 for an upgrade).

Since the current MAGIC FADCs have a resolution of 8 bit only, the analog signals are again split into two branches with a factor of 10 difference in gain. The low gain branch is delayed by 55 ns. The high gain branch is read out by the FADC using 15 time slices. In case the high gain pulse exceeds a set-able threshold, the low gain branch is then switched to the same FADC input and digitized into another 15 time samples. In case of a high gain signal below threshold no switch is performed and another 15 samples of the high gain wave form are read out. Therefore, for every event 30 time slices are stored. The FADC system can be read out with a maximum sustained rate of 1 kHz. A 512 kbytes FIFO memory allows short-time trigger rates of up to 50 kHz. The digital data is read out by a PC which saves it to a RAID system and a tape library. Tests of the complete read-out chain show that the achieved dynamic range (defined as the saturation signal charge divided by three times the RMS noise signal charge) is more than 1000 (Goebel et al. 2003).

From the recorded 30 time slices of the signal pulse, the signal intensity (charge) and arrival time are calculated off-line using an advanced pulse-reconstruction algorithm (see section 3.1). The arrival time information is of importance as Monte Carlo based simulations predict different time structures for $\gamma$-ray and hadron induced shower images as well as for images of single muons, see section 5.1. Moreover, the timing information may be used in the image cleaning to discriminate between camera pixels whose signals belong to the shower, and pixels which are affected by random background noise.

In order to further exploit the timing capabilities and the high bandwidth of the MAGIC read-out chain, the FADC system is being upgraded to a 2 GSamples/s sampling rate, using a bandwidth of 700 MHz and a resolution of 10 bit. In chapter 5 this DAQ upgrade project with an ultra-fast fiber-optic multiplexed FADC system is described, see also Bartko et al. (2005b).

### 2.2.7 The Calibration System

The purpose of the calibration system of the MAGIC telescope is to flat-field the camera response to the Cherenkov light from air showers, i.e. to have similar reconstructed charge signals in all pixels for the same photon illumination density, and to determine the absolute conversion factors between the reconstructed FADC signal and the Cherenkov light intensity. The calibration system consists of very fast (3-4 ns FWHM) and powerful ($10^8$-$10^{10}$ photons/ns/sr) light emitting diodes (NISHIA, single quantum well, Nakamura et al. (1995)) in three different wavelengths (370 nm, 460nm and 520 nm) and different intensities. These LEDs are installed in the middle of the mirror dish and illuminate the camera homogeneously. The light intensity is variable in the range of 4 to 700 photoelectrons per inner pixel. This enables to check the linearity of the read-out chain and to calibrate the whole dynamic range.

The absolute camera response to the calibration light flux may be obtained using a calibrated PIN (p-type, intrinsic, n-type) diode and three blinded camera pixels (input light attenuated by filters), which measure the absolute calibration light flux. For the calibration algorithms implemented see section 3.2.1. In addition, there are continuous light sources in four different colors and variable intensities to simulate the camera response to star and moon light. Figure 2.12 shows the elements of the MAGIC calibration system.

In dedicated calibration runs the MAGIC camera is illuminated by calibration LED light flashes and read out using the standard DAQ chain (Gaug et al. 2005). Moreover, using an external calibration trigger, it is possible to take calibration events interlaced with normal data taking at a rate of 50 Hz to correct for PMT gain variations on short time scales of a few minutes.

### 2.2.8 Monte-Carlo Simulations

The IACT method does not offer the possibility to evaluate the $\gamma$/hadron separation cut efficiency and the energy estimation by means of test beams of VHE $\gamma$-rays of known energy.

## 2.2 The Hardware Layout of the MAGIC Telescope

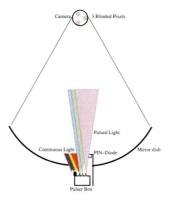

Figure 2.12: *Components of the MAGIC calibration system. Figure from Gaug (2006).*

Therefore, the operation of ground based Imaging Cherenkov telescopes requires a detailed Monte Carlo (MC) simulation of $\gamma$-ray and hadron-initiated air showers, as well as of the detector response. The Monte Carlo simulation for the MAGIC telescope (Majumdar et al. 2005) is divided into three stages: The development of $\gamma$- and hadron-initiated air showers is simulated with CORSIKA 6.019 (COsmic Ray SImulations for KAscade, Heck et al. (1998)) with some custom specific modifications (Sobczynska 2002). The electromagnetic part of the interactions is based on quantum electrodynamics (QED) calculations which are well under control. Major uncertainties, however, are introduced by the simulation of hadronic interactions, since the processes in air showers are dominated by low momentum transfers. Currently, these cannot be described by perturbative quantum chromo dynamics (pQCD). Therefore, the phenomenological model VENUS (Werner 1993) was used. The US standard atmosphere (Kneizys et al. 1996) was used in the simulations. For each Cherenkov photon arriving in a radius of 20 m around the telescope location the wavelength, arrival location and direction are stored a binary file.

The second stage of the simulation, the so-called *Reflector* (Moralejo 2002) program, accounts for the Cherenkov light absorption and scattering in the atmosphere, using the US standard atmosphere to compute the Rayleigh scattering plus the Elterman model (Elterman 1964) for the distribution of aerosols and ozone. Then the program performs the reflection of the surviving photons on the mirror dish (composed of 956 tiles) to obtain their location and arrival time on the camera plane.

Finally, the *camera* program (Blanch 2003) simulates the behavior of the MAGIC photomultiplier camera, trigger system and data acquisition electronics in a very detailed way. Realistic pulse shapes, noise levels and gain fluctuations obtained from the real MAGIC data have been implemented in the simulation. Furthermore, the overall light collection efficiency of the telescope has been tuned at the camera simulation level, using data from the comparison of the intensity of observed and simulated ring images from single muons at low impact parameters (Goebel et al. 2005).

To simulate a realistic shower mix from cosmic rays, protons and helium nuclei have been produced with energies between 30 GeV and 30 TeV following the measured spectra (Alcaraz et al. 2000a,b). The energy distribution of primary $\gamma$-rays was chosen to follow a pure power law with indexes of -2.6 and -1.0 to access also higher energies. The telescope pointing directions range up to 70 degrees in zenith angle and are evenly distributed in $\cos\theta$, with the directions of protons and Helium nuclei scattered isotropically within a 5° semiaperture cone around the telescope axis. Maximum impact parameters of 300 to 500 m have been simulated for $\gamma$-rays and nuclei, depending on the zenith angle.

## 2.3 Observations with the MAGIC Telescope

Observations with Cherenkov telescopes are generally conducted in moonless nights, due to the need for darkness to be able to detect the very faint flashes from Cherenkov light. With the MAGIC telescope also observations with partial moon are possible. Each observation night is typically split into time periods of up to a few hours duration during which the Cherenkov telescope tracks a given astrophysical target or position in the sky. Ideally, all other configuration parameters of the telescope are kept constant during this observation.

When observing a $\gamma$-ray source, the telescope not only records the images of $\gamma$-ray induced showers, but also the much more numerous background shower images due to the diffuse component of the cosmic ray flux (hadrons, electrons and photons) and LONS fluctuations. In order to separate this background from the source $\gamma$-rays, ON and OFF data are taken separately: For the ON data the telescope is directed exactly to the source, whereas for the OFF data the telescope is directed to a sky region nearby the object, but from where no $\gamma$-radiation is expected. The OFF-data can then be used to derive a background estimation for the ON-data. To reduce possible systematic differences between the ON and the OFF data, the OFF observations should be done under the same observation conditions (especially zenith angle, sky brightness, weather and so on). With this procedure a considerable part of the observation time is used for taking OFF data (Wittek 2001).

One may gain observation time as well as reduce the systematic differences between the ON and OFF data by producing ON and OFF data simultaneously. This is done in the so-called **wobble mode** (Fomin et al. 1994): The telescope is directed not exactly to the source position but to a point which is displaced from it by an angle $\Delta\beta$. The sign of $\Delta\beta$ is changed periodically (about every 20 minutes) in order to collect OFF data not only from the region on one side of the source. The two tracking positions corresponding to the two signs of $\Delta\beta$ are called wobble position 1 and wobble position 2. The ON data can now be obtained by analyzing the shower images with respect to the point in the camera which corresponds to the source position, the OFF data by analyzing the shower images with respect to some other point in the camera (the so-called "anti"-source position) which is not too close to the source position. As before, the source $\gamma$-rays are now essentially obtained by subtracting the OFF from the ON data. The basic assumption in this procedure is that the OFF data represent a good approximation of the background which is contained in the ON data.

The tracking offset $\Delta\beta$ should be chosen as a certain fraction of the diameter of the camera such that the source position is well inside the camera. Given an outer radius of the MAGIC camera of 2°, a radius of the inner part of 1.25° and a radius of the trigger region of 0.9°, a value of $\Delta\beta = \pm 0.4°$ is appropriate for the MAGIC telescope.

The technical implementation of these background determination techniques is discussed in section 3.5.

## 2.4 MAGIC II: The Second MAGIC Telescope

The MAGIC collaboration is currently constructing a second telescope (MAGIC II) at La Palma, see e.g. Teshima et al. (2005). Using both telescopes together pointing to the same object, the simultaneous observation of air showers with both MAGIC telescopes, will lead to a significant increase in sensitivity, see figure 3.40. The improved reconstruction of the air shower especially results in a better angular and energy resolution. The effective analysis energy threshold will be lowered. In addition, it improves the power to separate $\gamma$-ray showers from the backgrounds. In particular, single track events, e.g. muons, can effectively be rejected since they are mostly seen by only one telescope.

The MAGIC II telescope is located at a distance of 85 m from the MAGIC I telescope. Figure 2.13 shows the status of the installation of the second MAGIC telescope on La Palma. The structure of the MAGIC II telescope seen in the foreground has been installed in December 2005, without mirrors and camera. The commissioning of the MAGIC II telescope is planned for 2007. This will allow combined observations with the GLAST satellite telescope which observes in the energy range below 100 GeV and which will be launched in 2007. Simultaneous observations by MAGIC and GLAST will allow a precise cross calibration of both instruments (Bastieri et al. 2005), and will extend the energy spectrum of the combined observation to about 5 orders of magnitude (from about 100 MeV to more than 10 TeV).

MAGIC II will largely be a clone of the first MAGIC telescope in order to reduce time, manpower and money necessary for its construction. Nevertheless, several improvements will be introduced in MAGIC II:

- the camera will have a larger number (919) of small 0.1° pixels

- the trigger area will be increased by 72%, which increases the effective field of view

- the camera will have the option to be equipped with very high quantum efficiency GaAsP hybrid photo detectors (HPDs) (Hayashida et al. 2005)

- the complete signal processing chain from the light sensor to the digitizer is optimized for large bandwidth ultra-fast digitization (Antoranz et al. 2006).

The complete signal processing chain from the light sensor to the digitizer is optimized for large bandwidth. The pulses will be sampled and digitized by a 2 GSamples/s switched capacitor ring sampler (Turini et al. 2005). The shape of the 2 ns short Cherenkov light pulses emitted by $\gamma$-ray showers can thus be recorded with high resolution. This is expected to further improve the background rejection power due to the different time structure of hadron, muon and $\gamma$-ray events.

Figure 2.13: *Status of the installation of the second MAGIC telescope on La Palma in spring 2006: The structure of the MAGIC II telescope seen in the foreground has been installed in December 2005, without mirrors and camera. The MAGIC-I telescope seen in the background has been taking data since 2003.*

# Chapter 3
# Data Analysis

The observation of VHE $\gamma$-ray sources with the MAGIC telescope aims to study the following properties of these sources:

- the VHE $\gamma$-ray flux $\mathrm{d}N_\gamma/\mathrm{d}E\mathrm{d}A\mathrm{d}t$ as a function of the $\gamma$-ray energy
- the $\gamma$-ray source position and morphology (especially extension)
- the time variation of the $\gamma$-ray flux and source position/morphology.

The analysis of the MAGIC telescope data proceeds in four steps: The first step, see section 3.1, is the reconstruction of the number of photoelectrons (charge) and the arrival time of the Cherenkov signal for each pixel in the MAGIC camera. Thereafter, in section 3.2, each individual event is reconstructed yielding the estimated direction and energy of the primary particle as well as a measure of the probability to be a $\gamma$-ray or background event. In a third step (see section 3.3) the background and the $\gamma$-ray signal (number of excess events above background and significance) are determined from the different event samples (ON/OFF and source and background regions in the wobble data). In addition, a local VHE $\gamma$-ray sky map (3° × 3°) is calculated. In the last step (see section 3.4) the absolute $\gamma$-ray flux as a function of the $\gamma$-ray energy is determined. The analysis of the data taken in the wobble mode is described in section 3.5 and the achieved sensitivity compared to ON/OFF data. Section 3.6 summarizes the basic performance of the MAGIC telescope. Finally, in section 3.7, the sources for systematic uncertainties in the VHE $\gamma$-ray flux and in the source position measurements are discussed and the size of the errors are evaluated.

The data analysis presented in this thesis has been carried out in the framework of the standard MAGIC Analysis and Reconstruction Software MARS (Bretz & Wagner 2003). The algorithms and analysis procedures developed, as described in this section, have been implemented in this software package. The signal extraction algorithms and their performance presented in this thesis have been published by Bartko et al. (2005a); Albert et al. (2006e).

## 3.1 Charge/Arrival Time Extraction

To achieve a high sensitivity and a low analysis energy threshold, in each camera pixel the recorded signal of the Cherenkov light has to be accurately reconstructed, both in terms of the signal charge and the arrival time. High signal-to-noise ratio and signal reconstruction resolution are important, keeping the bias low at the same time.

In order to sample the $\sim 2.2$ ns FWHM wide PMT pulses with the 300 MSamples/s FADC system (see also section 2.2.6), the original pulse is folded with a stretching function of 6 ns leading to a FWHM greater than 6 ns. Due to the Nyquist theorem (Nyquist 1928) pulses to be digitized by a 300 MSamples/s FADC must be at least 6.6 ns long. On the other hand, the stretching washes out the different pulse time structures of the $\gamma$-ray and hadron showers (see section 5.1 and (Chitnis & Bhat 2001; Mirzoyan et al. 2006). Figure 3.2 shows an average of typical large signals.

This section is structured as follows: First in section 3.1.1 the characteristics of the MAGIC read-out system which are important for the signal reconstruction are presented. In section 3.1.2 the average pulse shapes for calibration pulses and Cherenkov light pulses ("cosmic pulses") are reconstructed from the recorded FADC samples. These pulse shapes are compared with the pulse shape used in the MC simulation. In section 3.1.3 the criteria for an optimal signal reconstruction are developed. In section 3.1.4 various signal reconstruction algorithms and their implementation in MARS are described. Thereafter, in sections 3.1.5 the different signal extraction algorithms are studied using Monte Carlo simulations. In sections 3.1.6 and 3.1.7 the performance of the standard MAGIC signal extraction algorithm is demonstrated for pedestal and calibration events. Section 3.1.8 discusses and summarizes the results.

### 3.1.1 Characteristics of the Current MAGIC Read-out System

For the signal reconstruction the following intrinsic characteristics of the current read-out system are of particular importance (see also section 2.2.6 for a more detailed description of the MAGIC read-out system):

- Inner and Outer pixels: The MAGIC camera (see section 2.2.4) has two types of pixels with the following differences:

  1. Size: The outer pixels have a factor four larger geometric area than the inner pixels (Barrio et al. 1998). Their (quantum-efficiency convoluted) effective area is about only a factor 2.6 larger because of a lower intrinsic quantum efficiency (QE) and a simpler Winston cone.

  2. Gain: The camera is flat-fielded in order to obtain similar reconstructed charge signals for the same photon illumination density in all pixels. To achieve this, the gain of the inner pixels has been adjusted to about a factor 2.6 higher than that of the outer ones (Gaug et al. 2004). This results in a lower effective noise charge from LONS for the outer pixels.

  3. Delay: The signal of the outer pixels is delayed by about 1.5 ns with respect to the inner ones due to a different size of the PMTs and high voltage settings.

## 3.1 Charge/Arrival Time Extraction

- Asynchronous trigger: The FADC clock cannot be synchronized with the trigger. Therefore, the time $\Delta t$ between the trigger decision and the first read-out sample is uniformly distributed in the range $\Delta t \in [0, T_{\text{FADC}}[$, where $T_{\text{FADC}} = 3.33$ ns is the digitization period of the MAGIC 300 MSamples/s FADCs.

- AC coupling: The PMT signals are AC-coupled at various places in the signal transmission chain. Thus the DC contribution to the PMT pulses from the light of the night sky is zero on average, only the signal fluctuations (root mean square, RMS) depend on the intensity of the LONS. In moonless nights, observing an extra-galactic source, a background rate of about 0.13 photoelectrons per nano-second per inner pixel has been measured, see e.g. Bartko et al. (2005b).

- Shaping: The optical receiver boards shape the pulse with a time constant of about 6 ns, much larger than the typical intrinsic pulse width (about 2 ns for a $\gamma$-ray shower). As the shaping time is larger than the width of a single FADC sample, strong correlation of the noise between neighboring FADC samples is expected.

- Instantaneous amplitudes: The MAGIC FADCs consist of a series of small comparators which measure the instantaneous amplitude of a pulse at a given time. No charge integration over the duration of a time sample is performed by the FADCs. Therefore, pulse structures are lost which have a frequency higher than the 300 MHz of the FADC sampling.

### 3.1.2 Pulse Shape Reconstruction

As mentioned above, the FADC clock is not synchronized with the trigger, therefore the time $\Delta t$ between the trigger decision and the first read-out sample is uniformly distributed in the range $\Delta t \in [0, T_{\text{FADC}}[$. $\Delta t$ can be determined (with a certain error caused by the photon arrival time spread) using the reconstructed arrival time $t_{\text{arrival}}$, the time difference between the first read-out FADC sample and a characteristic feature of the pulse like the position of the maximum, the center of gravity or of the half-maximum of the rising edge.

Figure 3.1a shows the raw FADC values as a function of the sample number for 1000 constant pulse generator pulses superimposed. For this study the response of the photomultipliers to Cherenkov light is simulated by a pulse generator which generates fast unipolar pulses of about 2.5 ns FWHM and a preset amplitude. These electrical pulses are transmitted using the same analog-optical link as for the PMT pulses and are fed to the MAGIC receiver board, where they are shaped with a time constant of about 6 ns. The pulse generator set-up is mainly used for test purposes of the receiver board, trigger logic and FADCs. Figure 3.1b shows the distribution of the corresponding reconstructed pulse arrival times. The distribution has a width of about 1 FADC period (3.33 ns). In principle, the time shift could be measured if one records the trigger signal itself in one additional FADC channel.

The reconstructed arrival times allow to determine an average pulse shape from the recorded signal samples: The recorded signal samples are shifted in time such that all signal pulses have their maximum at the same time. In addition, the signal samples are normalized event by event using the reconstructed charge of the pulse. Figure 3.2 shows

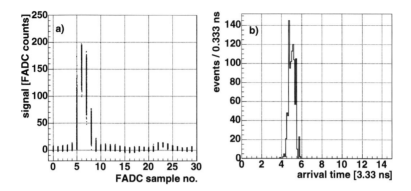

Figure 3.1: *a) Raw FADC samples of 1000 constant pulse generator pulses superimposed. b) Distribution of the reconstructed arrival time from the raw FADC samples shown in figure a). The width of the distribution is mostly due to the trigger jitter of 1 FADC period (3.33 ns).*

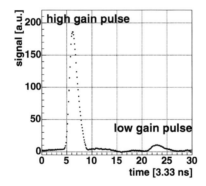

Figure 3.2: *Average reconstructed pulse shape from a standard fast pulse generator run (1000 events): The figure shows both the high gain and the low gain pulse. The FWHM of the high gain pulse is about 6.3 ns while the FWHM of the low gain pulse is about 10 ns (due to the pulse widening by the 55 ns lumped delay line).*

the averaged reconstructed signal of a fast pulser, corresponding to the raw FADC samples shown in figure 3.1a. The high and the low gain pulses are clearly visible. As already mentioned in section 2.2.6, the low gain pulse is smaller by a factor of about 10. It is

## 3.1 Charge/Arrival Time Extraction

delayed by about 55 ns with respect to the high gain pulse. The accuracy of the signal shape reconstruction depends on the accuracy of the arrival time and charge reconstruction. The relative statistical error of the value of every reconstructed point of the pulse shape is well below $10^{-2}$ while the systematic error is unknown.

Figure 3.3a shows the averaged, reconstructed pulse shapes for the generator pulses in the high and in the low gain branch, respectively, each normalized to an area of 1FADC count * 3.33 ns. The FWHM of the input pulses is about 2.5 ns. The FWHM of the average reconstructed high gain pulse shape is about 6.3 ns due to the pulse shaping, see section 2.2.6, while the FWHM of the average reconstructed low gain pulse shape is about 10 ns. The broadening of the low gain pulses with respect to the high gain pulses is due to the limited bandwidth of the passive 55 ns printed circuit board lumped delay line of the MAGIC receiver boards.

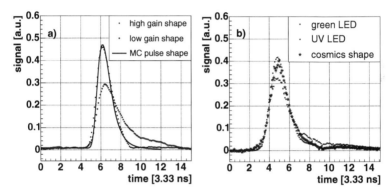

Figure 3.3: *a) Average reconstructed high gain and low gain pulse shapes from a fast pulse generator run. The FWHM of the low gain pulse is about 10 ns. The black line corresponds to the pulse shape implemented into the MC simulations (Blanch 2003). b) Average reconstructed high gain pulse shapes for green and ultra-violet (UV) calibration LED pulses as well as for air shower events. The FWHM of the UV calibration pulse and the air shower pulse are about 6.5 ns.*

Figure 3.3b shows the normalized, reconstructed pulse shapes, averaged over 1000 individual pulses, for green and ultra-violet (UV) calibration LED pulses as well as the normalized, reconstructed pulse shape for air shower events (again averaged over 1000 individual pulses). The pulse shape of the UV calibration pulses is quite similar to the reconstructed pulse shape for cosmic events, both have a FWHM of about 6.3 ns. The pulse shape for green calibration LED pulses is wider and has a pronounced tail. This is a feature of the LEDs used.

Since air showers from hadronic cosmic rays trigger the telescope much more frequently than $\gamma$-ray showers, the reconstructed pulse shape of the unselected cosmic events corresponds mainly to hadron induced showers. MC simulations show that electromagnetic shower light flashes have a typical FWHM of 2 ns and hadronic light flashes a FWHM of about 4 ns (see e.g. Chitnis & Bhat (2001), Mirzoyan et al. (2006) and figure 5.1). The applied pulse shaping (necessary for the 300 MSamples/s FADC read-out) washes out these differences.

### 3.1.3 Criteria for an Optimal Signal Extraction

The goal for an optimal signal reconstruction algorithm is to compute an unbiased estimate of the charge and arrival time of the Cherenkov signal with the smallest possible error of the reconstructed charge for all signal intensities. The MAGIC telescope design has been optimized for a low energy threshold of observation, so the signal reconstruction of small signals is of particular importance.

An accurate determination of the signal arrival time may help to distinguish between signal and LONS background: The signal arrival times vary smoothly from pixel to pixel while the background noise is randomly distributed in time. Therefore, one has to develop a signal extraction algorithm which reconstructs both the signal charge and arrival time.

### 3.1.4 Signal Extraction Algorithms

There are four algorithms implemented for the reconstruction of the signal charge and the arrival time in the MAGIC Analysis and Reconstruction Software MARS (Bretz & Wagner 2003):

- Fixed Window algorithm
- Sliding Window algorithm with Amplitude-Weighted Time
- Cubic Spline with Sliding Window or Amplitude Extraction
- Digital Filter.

#### 3.1.4.1 Fixed Window Extraction Algorithm

This signal extraction algorithm simply adds the pedestal-subtracted FADC slice contents of consecutive FADC slices within a fixed range (window). The window has to be chosen large enough (at least 6 FADC slices in the high-gain corresponding to 20 ns, see Albert et al. (2006e)) to always cover the complete pulse, otherwise the inevitable jitter in the pulse position with respect to the FADC slice numbering would lead to an incomplete integration of the pulse. For this reason, the fixed window algorithm adds up more noise than the other considered signal reconstruction algorithms. Due to the AC-coupling of the read-out chain the average FADC slice content is zero in absence of a signal. Therefore, the reconstructed signals have no bias.

In the current implementation, the fixed window algorithm does not calculate arrival times but assumes that all signals have the same arrival time.

## 3.1 Charge/Arrival Time Extraction

### 3.1.4.2 Sliding Window Extraction Algorithm with Amplitude-Weighted Time

This signal extraction algorithm searches for the maximum integral content among all possible FADC windows of fixed length (typically 4 FADC slices for the high-gain and 6 FADC slices for the low gain) contained in a defined larger time range (global window) of typically 14 FADC slices. The arrival time is calculated from the window with the largest integral as:

$$t = \frac{\sum_{i=i_0}^{i_0+ws-1} s_i \cdot t_i}{\sum_{i=i_0}^{i_0+ws-1} s_i} \quad , \tag{3.1}$$

where $i$ denotes the FADC slice index, starting from slice $i_0$ and running over a window of size $ws$. The $s_i$ are the pedestal-subtracted FADC slice contents and the $t_i$ are the corresponding times relative to the first recorded FADC slice.

### 3.1.4.3 Cubic Spline with Sliding Window or Amplitude Extraction

This signal extraction algorithm interpolates the pedestal subtracted FADC sample contents using a cubic spline algorithm, adapted from Press et al. (2002). In a second step, it searches for the position of the spline maximum. Thereafter, two possibilities are offered for the charge reconstruction:

1. **Amplitude**: the amplitude of the spline maximum is taken as the reconstructed signal.

2. **Integral**: The spline is integrated in a window of fixed size (typically 3 FADC slices, corresponding to 10 ns), with integration limits determined with respect to the position of the spline maximum.

The pulse arrival time can be computed in two ways:

1. **Time of the pulse maximum**: The position of the spline maximum determines the arrival time.

2. **Time of the pulse half maximum**: The position of the half maximum at the rising edge of the pulse determines the arrival time.

The pulse FWHM of 2 ns for Cherenkov pulses from $\gamma$-ray showers is small compared to the electronic signal shaping time of about 6 ns. Therefore, for $\gamma$-ray signals the pulse amplitudes and integrals are to a good approximation proportional to each other. As the signal for hadron showers may be wider (about 4 ns FWHM) also the pulse widths after shaping of hadron showers may be slightly wider than the ones for $\gamma$-ray showers. However, this difference is small, such that only a marginal improvement in the $\gamma$/hadron separation may be expected.

For varying pulse shapes, the pulse amplitudes are no longer proportional to the pulse integral. In this case the pulse integral is the correct measure of the total signal charge (number of photoelectrons). In principle, the ratio between pulse amplitude and pulse integral may yield a contribution to the $\gamma$/hadron separation. See section 5.1 for further discussions for the case without pulse shaping.

### 3.1.4.4 Digital Filter Extraction Algorithm

The goal of the digital filtering method (Cleland & Stern 1994; Papoulis 1977) is to optimally reconstruct from the FADC samples the amplitude and arrival time of a signal whose shape is known. With the digital filter extraction algorithm, the noise contributions to the amplitude and arrival time reconstruction are minimized, see also Bartko et al. (2005a).

For the digital filtering method to work properly, two conditions have to be fulfilled:

- The normalized signal shape has to be constant.

- The noise properties must be constant, i.e. the noise level should not vary with time and the noise should be independent of the signal amplitude.

As the pulse shape is mainly determined by the artificial pulse stretching on the optical receiver board, the first assumption holds to a good approximation for all pulses with intrinsic signal widths smaller than the shaping constant, e.g. for $\gamma$-ray showers. Hadronic showers may lead to wider pulse shapes which causes differences between the reconstructed and the true signal charge and arrival time. However, hadronic showers are background and can be efficiently rejected e.g. using the random forest method, see section 3.2.5. Also the second assumption is fulfilled to a good approximation: Signal and noise are independent and the measured pulse is a linear superposition of the signal and noise contributions.

The digital filter method performs a numerical fit (by means of minimizing a $\chi^2$) of the assumed signal shape (with the two free parameters arrival time and pulse amplitude) to the measured FADC samples. Through the fit the full noise auto-correlation is taken into account: Let $g(t)$ be the assumed and normalized signal shape (from figure 3.2), $E$ the signal amplitude and $\tau$ the time shift between the physical signal and the assumed signal shape. Then the time dependence of the signal, $y(t)$, is given by

$$y(t) = E \cdot g(t - \tau) + b(t) , \qquad (3.2)$$

where $b(t)$ is the time-dependent noise contribution. For small arrival time shifts $\tau$ (usually smaller than one FADC sample width), the time dependence can be linearized. Discrete measurements $y_i$ of the signal at times $t_i$ ($i = 1, ..., n$) have the form

$$y_i = E \cdot g_i - E\tau \cdot \dot{g}_i + O(\tau^2) + b_i , \qquad (3.3)$$

where $\dot{g}(t)$ is the time derivative of the signal shape and $O(\tau^2)$ is a term proportional to $\tau^2$ which vanishes for vanishing values of $\tau$.

The correlation of the noise contributions at times $t_i$ and $t_j$ can be expressed by the noise autocorrelation matrix $\boldsymbol{B}$:

$$B_{ij} = \langle b_i b_j \rangle - \langle b_i \rangle \langle b_j \rangle . \qquad (3.4)$$

Figures 3.4a,b show the noise autocorrelation matrix for an open camera and low/high LONS, respectively. The noise is dominated by LONS pulses shaped with 6 ns time constant. The ratio between the two noise autocorrelation matrices a) and b) is presented in figure 3.4c. The ratio is not a constant, because the entries of $\boldsymbol{B}$ do not simply scale with

## 3.1 Charge/Arrival Time Extraction

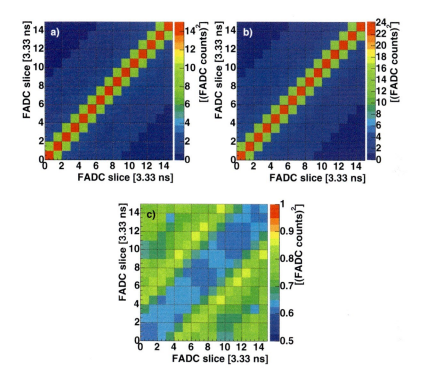

Figure 3.4: *Noise autocorrelation matrix **B** for an open camera and averaged over all pixels: a) the telescope pointing off the galactic plane (low LONS fluctuations) b) the telescope pointing into the galactic plane (high LONS). c) The ratio between a) and b). One can see that the entries of **B** do not simply scale with the amount of LONS.*

the amount of LONS. The noise is a superposition of a constant electronics noise and noise due to LONS.

For a given pulse, $E$ and $E\tau$ can be estimated from the $n$ FADC measurements $\boldsymbol{y} = (y_1, ..., y_n)$ by minimizing the deviation between the measured and the known pulse shape, and taking into account the known noise auto-correlation, i.e. minimizing the following expression (in matrix form):

$$\chi^2(E, E\tau) = (\boldsymbol{y} - E\boldsymbol{g} - E\tau\dot{\boldsymbol{g}})^T \boldsymbol{B}^{-1} (\boldsymbol{y} - E\boldsymbol{g} - E\tau\dot{\boldsymbol{g}}) + O(\tau^2) \ . \tag{3.5}$$

This leads to the following solution for $\overline{E}$ and $\overline{E\tau}$:

$$\overline{E} = \boldsymbol{w}_{\text{amp}}^T(\Delta t)\boldsymbol{y} + O(\tau^2) \quad , \quad \boldsymbol{w}_{\text{amp}}(\Delta t) = \frac{(\dot{\boldsymbol{g}}^T \boldsymbol{B}^{-1} \dot{\boldsymbol{g}}) \boldsymbol{B}^{-1}\boldsymbol{g} - (\boldsymbol{g}^T \boldsymbol{B}^{-1} \dot{\boldsymbol{g}}) \boldsymbol{B}^{-1}\dot{\boldsymbol{g}}}{(\boldsymbol{g}^T \boldsymbol{B}^{-1} \boldsymbol{g})(\dot{\boldsymbol{g}}^T \boldsymbol{B}^{-1} \dot{\boldsymbol{g}}) - (\dot{\boldsymbol{g}}^T \boldsymbol{B}^{-1} \boldsymbol{g})^2} \ , \tag{3.6}$$

$$\overline{E\tau} = \boldsymbol{w}_{\text{time}}^T(\Delta t)\boldsymbol{y} + O(\tau^2) \quad , \quad \boldsymbol{w}_{\text{time}}(\Delta t) = \frac{(\boldsymbol{g}^T \boldsymbol{B}^{-1} \boldsymbol{g}) \boldsymbol{B}^{-1}\dot{\boldsymbol{g}} - (\boldsymbol{g}^T \boldsymbol{B}^{-1} \dot{\boldsymbol{g}}) \boldsymbol{B}^{-1}\boldsymbol{g}}{(\boldsymbol{g}^T \boldsymbol{B}^{-1} \boldsymbol{g})(\dot{\boldsymbol{g}}^T \boldsymbol{B}^{-1} \dot{\boldsymbol{g}}) - (\dot{\boldsymbol{g}}^T \boldsymbol{B}^{-1} \boldsymbol{g})^2} \ , \tag{3.7}$$

where $\Delta t$ is the time difference between the trigger decision and the first read-out sample, see section 3.1.2. Thus $\overline{E}$ and $\overline{E\tau}$ are given by a weighted sum of the discrete measurements $y_i$ with the weights for the amplitude, $w_{\text{amp}}(\Delta t)$, and time shift, $w_{\text{time}}(\Delta t)$, up to terms of $O(\tau^2)$. To reduce the error term $O(\tau^2)$ the fit can be iterated using $g(t_1 = t - \tau)$ and the weights $w_{\text{amp/time}}(\Delta t + \tau)$ (Cleland & Stern 1994). Figure 3.5 shows examples of digital filter weights for MC simulated $\gamma$-ray pulses and measured cosmic pulses for both the high and the low gain pulse. The black lines represent the normalized signal shapes $g(t)$ (multiplied with 5 for better visibility), the blue lines the amplitude weights $w_{\text{amp}}(t)$, and the red lines the time weights $w_{\text{time}}(t)$. Note, that for the low gain the same weights are used to reconstruct the MC simulated pulses and the cosmic pulses, because the noise autocorrelation in the low gain cannot be measured. In principle, it could be calculated or determined from MC simulations.

The expected contributions of the noise to the error of the estimated amplitude and timing only depend on the the shape $g(t)$ and the noise auto-correlation $\boldsymbol{B}$. The corresponding analytic expressions can be found in Cleland & Stern (1994).

### 3.1.5 Monte Carlo Studies of Signal Extraction

Some characteristics of the signal extraction algorithm can only be investigated by means of Monte Carlo simulations of signal pulses and noise (for the MAGIC MC simulations see Majumdar et al. (2005) and section 2.2.8). While under real conditions the number of photoelectrons is randomly distributed according to a Poission distribution due to the PMT photoelectron statistics, simulated pulses of a specific number of photoelectrons can be generated. Moreover, the same pulse can be studied with or without added noise and the noise level can be varied. For the subsequent studies, the following settings have been used:

## 3.1 Charge/Arrival Time Extraction

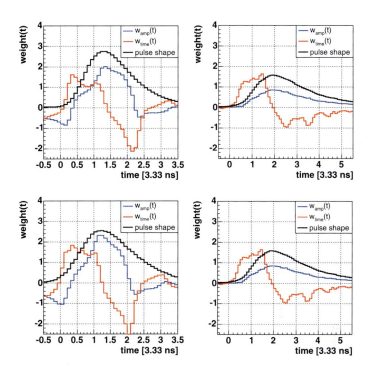

Figure 3.5: *Examples of digital filter weights. Top: MC simulation pulses, bottom: cosmic pulses. On the left side the digital filter weights for the high gain pulses are shown, on the right side, the ones for the low gain. The black lines present the normalized signal shapes $g(t)$ (multiplied with a factor of 5 for better visibility), blue lines show the amplitude weights $w_{\mathrm{amp}}(t)$, and red lines the time weights $w_{\mathrm{time}}(t)$.*

- The conversion factor from photoelectrons to integrated charge over the whole pulse was set to 7.8 FADC counts per photoelectron, there are no gain fluctuations.
- The relative timing between the trigger and the signal pulse was uniformly distributed over 1 FADC slice.
- The LONS has been simulated approximately as in extra-galactic source observation conditions.
- The total dynamic range of the entire signal transmission chain was set to infinite, thus the detector has been simulated to be completely linear.
- The intrinsic arrival time spread of the photons was set to be 1 ns, as expected for $\gamma$-ray showers.
- No PMT time spread (negligible compared to the shaping time of 6 ns) has been simulated.
- Only one inner pixel has been simulated.

Figure 3.6 depicts the signal pulse shape of a typical MC event (for the chosen parameterization of the pulse shape see Blanch (2003)) together with the simulated FADC samples. The FADC measurements are affected by noise, e.g. at t=2*3.33ns there is a random noise peak due to LONS. The digital filter has been applied to reconstruct the signal size and timing. Using this information together with the assumed MC pulse shape, the pulse shape is reconstructed and shown as well. It agrees well with the simulated shape (on average the $\chi^2$ value equals the number of degrees of freedom).

Below, Monte Carlo simulations are used to determine especially the following quantities for each of the tested signal extraction algorithms:

- The charge resolution as a function of the input signal charge.
- The charge extraction bias as a function of the input signal charge.
- The time resolution as a function of the input signal charge.

Figure 3.7 presents a) the charge and b) the arrival time resolution as a function of the input pulse charge for MC simulations assuming an extra-galactic background for different signal extraction algorithms. The charge and time resolutions are roughly proportional to the pedestal RMS level (square root of number of LONS photoelectrons per time). The digital filter yields the best charge and timing resolution of the studied algorithms (Bartko et al. 2005a). Although the charge resolution of the digital filter is expected to be independent of the charge, it increases slightly with increasing charge. This is due to residual event to event differences between the actual pulse form and the assumed signal shape.

Figure 3.8 shows the bias of the reconstructed charge as a function of the simulated input charge. All three studied signal extraction algorithms show a bias for small signals. This is due to the search for the maximum signal in the integration window. In case of a

## 3.1 Charge/Arrival Time Extraction

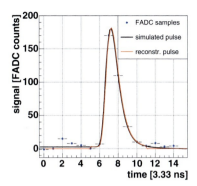

Figure 3.6: *Simulated signal pulse shape (black line) and FADC samples (blue points) for a typical MC event. The FADC measurements are affected by noise, e.g. at t=2\*3.33ns there is a noise peak due to LONS. The red line shows the reconstructed pulse shape using the digital filter result for the charge and arrival time and the assumed MC pulse shape. The simulated and reconstructed pulses agree well.*

very low input charge the algorithms reconstruct the highest charge signal from electronics noise or LONS. Above the image cleaning threshold of 5 photoelectrons the digital filter and the spline signal extraction algorithms show biases below 0.1 photoelectrons (below 2% of the input charge). The sliding window has a somewhat larger bias.

In the case of a large background light level (e.g. a star is imaged in the camera pixel) the charge and time resolution is reduced. Nevertheless, above the threshold of about 5 photoelectrons the digital filter and the spline signal extraction algorithm show very low biases.

### 3.1.6 Pedestal Reconstruction

The pedestal is the average FADC count content for the signal baseline (no input signal). In the DAQ system it is set to a value of around 15 FADC counts. To determine the pedestal setting off-line, dedicated pedestal runs are used, where the MAGIC read-out is randomly triggered.

For small signals there is no switch of the DAQ chain from the high to the low gain for the second 15 FADC samples out of the total recorded 30 FADC samples, see section 2.2.6. In this case the contents of the second 15 FADC samples are used to calculate the pedestal (which is defined as the average of these 15 FADC samples). After every 500 such events, for which no switch from the high to the low gain happens, the pedestal value is updated.

The fluctuations of the signal baseline are due to electronics noise as well as LONS fluctuations. Thus the pedestal RMS is a measure for the total integrated noise. Fig-

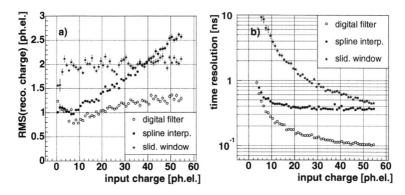

Figure 3.7: a) Charge and b) arrival time resolution as a function of the input pulse charge for MC simulations for the Digital Filter, a cubic spline interpolation (charge= spline integral over 1.5 FADC samples around the pulse maximum, time=half maximum position of the rising edge of the pulse) and a sliding window of 6 FADC samples (charge=samples sum, time = pulse barycenter) (Bartko et al. 2005a).

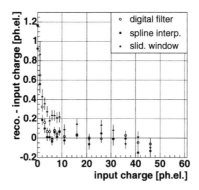

Figure 3.8: Bias of the reconstructed charge as a function of the simulated input charge. Above the image cleaning threshold of 5 photoelectrons the digital filter and the spline signal extraction algorithms show biases below 0.1 photoelectrons (below 2% of the input charge). The sliding window shows a somewhat larger bias.

## 3.1 Charge/Arrival Time Extraction

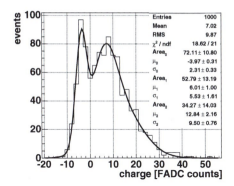

Figure 3.9: *Distribution of the extracted charge from a pedestal run (galactic star background) for a typical pixel (No. 100) using the digital filter. One can see a two peak structure. The left peak is due to the signal baseline and electronics noise. The right peak is due to one or more photoelectrons from LONS. The average reconstructed signal is above zero due to the search feature of the digital filter (like the sliding window and the spline interpolation algorithms) for the largest signal above the baseline, the bias (see e.g. figure 3.8 above).*

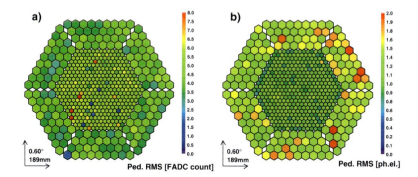

Figure 3.10: *a) RMS of the charge distribution of the digital filer applied to pedestal events in units of FADC counts. b) The same data after calibration to photoelectrons.*

ure 3.9 shows the distribution of the extracted charge from a pedestal run (galactic star background) for a typical pixel (No. 100) using the digital filter. One can see a two peak structure. The left peak is due to the signal baseline and electronics noise. The right peak is due to one or more photoelectrons from LONS. The relative size of the two peaks depends on the LONS level, in case of a closed camera there is just one peak around zero reconstructed signal. The average reconstructed signal in the presence of LONS is above zero. This is due to the search feature of the digital filter (like the sliding window and the spline interpolation algorithms) for the largest signal above the baseline, the bias (see e.g. figure 3.8 above). Due to the search for the largest signal the charge distribution cannot be easily used to compute the calibration conversion factor between the charge in FADC counts and photoelectrons.

Figure 3.10a depicts the RMS of the charge distribution of the digital filter applied to pedestal events in units of FADC counts for all pixels in the MAGIC camera. Figure 3.10b shows the same data after calibration to photoelectrons. The average pedestal RMS is about 1.1 photoelectrons for inner pixels and about 1.5 photoelectrons for outer pixels.

### 3.1.7 Calibration Pulse Reconstruction

The digital filter has been applied to events when the camera is illuminated with the very fast calibration light pulser, see section 2.2.7. The digital filter yields the reconstructed charge in units of FADC counts. Figure 3.11a presents the average reconstructed charge in units of FADC counts as a function of the average reconstructed charge in photoelectrons. The latter is obtained by the F-Factor method described in section 3.2.1. Up to signals of a few hundred photoelectrons the reconstruction is linear within errors. Figure 3.11b shows the measured timing resolution using the Digital Filter for different calibration LED pulses as a function of the mean reconstructed pulse charge, see also section 3.2.1. For signals of 10 photoelectrons the timing resolution is as good as 700 ps, for very large signals a timing resolution of about 200 ps can be achieved. This timing resolution is dominated by the intrinsic width of the calibration light pulse (about 2 ns) and the transit time spread of the PMT of several hundred ps (not included in the Monte Carlo simulations). The intrinsic resolution of the timing extraction algorithm is less.

### 3.1.8 Performance and Discussion

For known constant signal shapes and noise auto-correlations the digital filter yields the best theoretically achievable signal and timing resolution for $\gamma$-ray showers. Due to the pulse shaping of the Cherenkov signals the algorithm can be applied to reconstruct their charge and arrival time, although there are some fluctuations of the pulse shape and noise behavior. The digital filter reduces the noise contribution to the error of the reconstructed signal. Thus it is possible to lower the image cleaning levels and the analysis energy threshold (Bartko et al. 2005a). The timing resolution is as good as a few hundred ps for large signals.

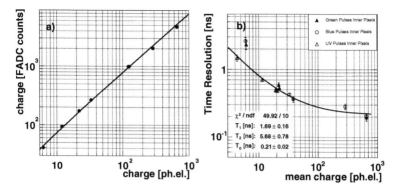

Figure 3.11: *a) Linearity of the signal reconstruction using the digital filter: Average reconstructed charge in units of FADC counts as a function of the average reconstructed charge in photoelectrons. Up to signals of a few hundred photoelectrons the reconstruction is linear within errors. b) Arrival time resolution using the Digital Filter as a function of the input pulse charge for different calibration LED pulses. The full line shows a parameterization of the time resolution, see section 3.2.1.*

## 3.2 Event Reconstruction

After the signal (in units of FADC counts) and the arrival time have been reconstructed for each individual pixel of the MAGIC camera, the event properties are reconstructed from the shower image in the camera. In a first step (section 3.2.1) calibration constants are applied to the reconstructed signals to account for gain and arrival time differences of the different camera pixels (Gaug et al. 2005). Defect pixels are interpolated by the average signal of the adjacent pixels (section 3.2.2). Thereafter, in section 3.2.3, a so-called image cleaning is performed which rejects pixels with a low signal-to-noise ratio. In section 3.2.4 the remaining ("cleaned") shower images are characterized by image parameters (Hillas 1985). Finally, for each event a measure for the probability to be a background event, the energy and the arrival direction of the primary particle are calculated, sections 3.2.5 to 3.2.7.

### 3.2.1 Calibrations

The MAGIC telescope PMT camera requires precise and regular calibration over a large dynamic range (Gaug et al. 2005). The calibration provides the conversion constants from the extracted signal charge in units of FADC counts and arrival time with respect to a particular FADC clock tick to the physical quantities of signal charge in photoelectrons (or Cherenkov light pulse intensity in photons) and absolute signal timing. For this purpose an optical calibration system (Schweizer et al. 2002) consisting of a number of ultrafast and powerful LED pulsers is used, which illuminate the MAGIC camera homogeneously (see section 2.2.7). There are four methods to determine the calibration constants for the charge:

1. Excess noise factor method (calibration ADC counts - photoelectrons)
2. Blind pixel method (calibration ADC counts - photoelectrons)
3. PIN diode method (calibration ADC counts - photos arriving on camera)
4. Muon ring calibration (calibration ADC counts - photos arriving on mirror).

#### 3.2.1.1 Excess Noise Factor Method ("F-Factor Method")

For this method one assumes that the number of photons impinging on the photocathode has a Poisson variance, that the photon detection efficiency is independent of the place where and under which angle the photoelectron is released and that the excess noise introduced by the gain fluctuations does not depend on the signal amplitude. From this one can derive (Mirzoyan & Lorenz 1997) the following relation for the average number of photoelectrons due to the calibration pulses, $N_{\text{ph.el.}}$:

$$N_{\text{ph.el.}} = F^2 \frac{\mu^2}{\sigma_1^2 - \sigma_0^2} , \qquad (3.8)$$

where $\sigma_0$ describes the error of the reconstructed charge due to LONS fluctuations and the signal extraction algorithm intrinsic uncertainties, see section 3.1.6, $\sigma_1$ is the measured standard deviation of the reconstructed charge and $\mu$ is mean reconstructed charge. $\mu$, $\sigma_0$ and $\sigma_1$ are measured in units of FADC counts. $F$ denotes the so-called excess noise factor, previously measured in the laboratory. The excess noise factor method yields one value of $N_{\text{ph.el.}}$ per pixel for an LED calibration run.

#### 3.2.1.2 Calibration to Equivalent Photoelectrons

The camera of the MAGIC telescope consists of 397 inner pixels and 179 outer pixels, which have four times the area of an inner pixel. In an ideal case all pixels would have the same photon to photoelectron conversion efficiency. Nevertheless, there are considerable differences in this efficiency. Especially the outer pixels are affected by somewhat larger losses of the light collection and photon conversion efficiency due to the selection of the PMTs with the highest QE as inner pixels and a somewhat simpler design of the light guides for the outer pixels (Barrio et al. 1998).

To account for these differences in photon to photoelectron conversion efficiency, the data are calibrated to *equivalent photoelectrons*. The number of equivalent photoelectrons is proportional to the photon fluence per pixel and does not reflect the actual number of photoelectrons emitted by the photocathode of the PMT.

The median number of photoelectrons in a calibration run of the inner pixels defines the reference value $\langle N_{\text{ph.el.}} \rangle$ for the calibration. All reconstructed charges $\mu_{\text{data}}^i$ from the

## 3.2 Event Reconstruction

data are then multiplied with a conversion factor:

$$c^i_{\text{phe}} = \langle N_{\text{ph.el.}} \rangle \frac{R^i_{\text{area}}}{\mu^i_{\text{calib}}}, \quad (3.9)$$

where $i$ is the pixel index, $\mu^i_{\text{calib}}$ is the mean charge in FADC counts from the calibration pulses in pixel $i$ and $R^i_{\text{area}}$ is a measure of the pixel area ($R^i = 1$ for all inner pixels and $R^i = 4$ for all outer pixels). All values of photoelectrons in the following refer to *equivalent photoelectrons*. Figure 3.12a shows the conversion factors $c^i_{\text{phe}}$ between the reconstructed signal charge in FADC counts and equivalent photoelectrons determined from one calibration run. The average conversion factor for the inner pixels is about 0.15 photoelectrons per FADC count for the inner pixels and about 0.55 photoelectrons per FADC count for the outer pixels.

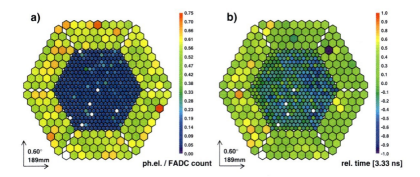

Figure 3.12: *a) Conversion factors $c^i_{\text{phe}}$ between the reconstructed signal charge in FADC counts and equivalent photoelectrons, b) Calibration signal relative arrival time for a 10 LED UV calibration run in units of FADC sampling intervals (3.33 ns). White pixels denote defect pixels.*

### 3.2.1.3 PIN Diode Cross-Calibration Method

Each pixel is calibrated using the F-Factor Method with light pulses of three different wavelengths. However, this just gives the calibration factor between the reconstructed charge in FADC counts and the number of photoelectrons released by the light input signal to the camera pixel. In addition, one has to calibrate the photon flux hitting the pixels.

An absolute calibration between FADC counts and light input signal photons can be obtained by comparing the signal of the pixels with the one obtained from a laboratory calibrated 1 cm² PIN diode. The experimental setup is shown in figure 2.12. For further information, see Gaug et al. (2005). The systematic error is up to 8%. The PIN diode method is not yet routinely used to calibrate the data.

#### 3.2.1.4 Blinded Pixel Method

Three selected 0.1° pixel PMTs with accurately measured quantum efficiencies are installed in addition in the camera. They are "blinded" with filters of known transmission in order to reduce the number of photoelectrons per calibration pulse in a controlled way, such that on average only for one in 10-20 light pulses a single photoelectron is detected. The number of photoelectrons follows a Poisson distribution, which allows to determine the average number of photoelectrons from the measured pulse charge distribution. Using the known filter transparency and quantum efficiency one can determine the absolute calibration light flux, see e.g. Gaug (2006). The blinded pixel method is not yet routinely used to calibrate the data.

#### 3.2.1.5 Muon ring calibration

The absolute overall light collection efficiency of the MAGIC telescope can be calibrated using isolated muons hitting the reflector (Goebel et al. 2005; Rose 1995). The geometry and the energy of the muons are reconstructed from the measured ring images and compared with Monte Carlo predictions. The amount of Cherenkov light produced by muons can be modeled with small systematic uncertainties. Muon images are recorded during normal observations with a rate of about 2 Hz. A continuous calibration can therefore be performed with no need for dedicated calibration runs. The parameters of the MC simulations (especially the reflectivity of the mirror) are adjusted yielding an absolute calibration of the reconstructed energy of the observed $\gamma$-ray showers. In addition, the width of the muon ring images can be used to monitor the spot size of a point-like source in the camera during normal data taking. It agrees well with the point spread function measured from star images, see section 3.6.2.

#### 3.2.1.6 Time Calibration

The photomultipliers introduce a time delay, the transit time (TT), in the amplified photoelectrons signal, depending on the applied high-voltage (HV), typically 12 ns for the ET 9116 PMTs and 18 ns for the 9117 PMTs at the nominal HV of 1400 V. Together with smaller relative delays due to different lengths of the optical fibers, these delays have to be calibrated relative to each other in order to obtain a correct timing information for the analysis (Gaug et al. 2005).

Using the light pulser at different intensities, the time offsets and time spreads of the readout and detection chain are measured. Event by event, the reconstructed arrival time difference of every channel with respect to a reference channel was measured and its mean and RMS calculated. The former yields the measured relative time offset while the latter is the convolution of the arrival time resolutions of the measured and the reference channel.

Figure 3.11b shows the time resolution (RMS of the arrival time differences histogram, divided by the square root of 2), measured at different intensities. The measurements have

## 3.2 Event Reconstruction

been fitted by the following ansatz (black line in figure 3.11b):

$$\Delta T_{\text{cosmics}} \approx \sqrt{\frac{4\,\text{ns}^2}{<Q>/\text{ph.el.}} + \frac{4\,\text{ns}^2}{<Q>^2/\text{ph.el.}^2} + 0.04\,\text{ns}^2} \,. \quad (3.10)$$

This time resolution still contains the contribution due to the calibration light pulse width of about 2 ns and is therefore an upper bound to the achievable timing resolution for Cherenkov light events.

Figure 3.12b shows the calibration signal relative arrival time (with respect to pixel 100) for a 10 LED UV calibration run in units of FADC sampling intervals (3.33 ns). There are relative arrival time differences up to a few ns. As the arrival time can be determined with a resolution below 1 ns, the relative arrival time differences must be corrected for.

### 3.2.2 Bad Pixel Treatment

A few camera pixels cannot be calibrated due to various reasons, mostly hardware failure or the image of a very bright star. In case a non-calibrated pixel has at least three calibrated neighbors, the charge after calibration is set for this non-calibrated pixel equal to the average calibrated charge of the neighboring pixels (Tonello 2006). Up to 5% of the pixels cannot be calibrated and are interpolated.

### 3.2.3 Image Cleaning

In order to reject pixels with a low signal-to-noise ratio which may influence the subsequent parameterization of the shower images, so-called absolute image cleaning tail cuts are applied in a two step procedure: All camera pixels with a charge of at least 10 photoelectrons, that have a neighboring pixel with a charge of at least 10 photoelectrons, are assigned to be so-called "core pixels" of the shower image. All pixels that are no core pixels but have a charge signal of at least 5 photoelectrons and are neighbor to a core pixel are assigned to be "boundary pixels". All inner camera pixels that are neither core nor boundary pixels are set to zero signal. These tail cuts are accordingly scaled by a factor of four for the larger size of the outer pixels of the MAGIC camera, see e.g. Wittek (2002c). By applying the image cleaning algorithm one artificially raises the analysis energy threshold and one loses the information contained in the shower tail. Especially for low energy showers, this decreases the $\gamma$/hadron separation power of the cleaned images.

Figure 3.13a depicts the camera image of a $\gamma$-ray candidate event. The color scale denotes the calibrated pixel signal in photoelectrons. In figure 3.13b the color scale illustrates the reconstructed and calibrated pulse arrival time for all pixels. The pixels contained in the shower image show similar arrival times. Figure 3.13c shows the same event after the application of the image cleaning algorithm. The green ellipse was fitted to the brightness distribution (pixel charge per pixel area) after image cleaning. The red lines show the major and minor axis of this ellipse.

The arrival time of the charge pulse in each pixel can also be used in the image cleaning. According to MC simulations the arrival time of the Cherenkov photons from the shower is expected to vary smoothly over the shower image. Contrary to that the arrival time of

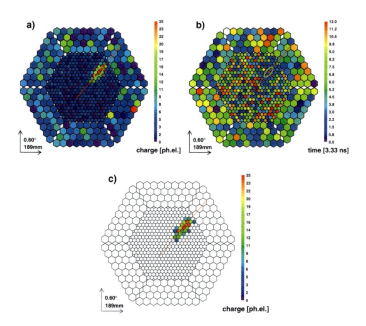

Figure 3.13: *Shower picture before and after image cleaning of a $\gamma$-ray candidate event: a) calibrated signal in equivalent photoelectrons b) calibrated arrival time in FADC sampling intervals (3.33 ns) c) calibrated signal after application of the image cleaning algorithm. The green ellipse was fitted to the brightness distribution after image cleaning. The red lines show the major and minor axis of this ellipse.*

noise pulses from LONS fluctuations are randomly distributed in time. Presently, there are several timing based image cleaning algorithms under study and evaluation. They show promising results, see e.g. Gaug (2006).

### 3.2.4 Image Parameterization / Hillas Parameters

For further analysis the cleaned shower pictures obtained with the MAGIC PMT camera are characterized by a set of image parameters, first proposed by Hillas (1985), see also Wittek (2002b). The basis of this image parameterization is that the Cherenkov light distribution for a $\gamma$-ray induced shower in the camera is in a first approximation elliptical. This ellipse can be characterized by the moments of the light intensity distribution. The Hillas parameters correspond to brightness, position, orientation and shape of the image in the camera. Figure 3.14 illustrates the definition of the basic Hillas parameters *Width*,

## 3.2 Event Reconstruction

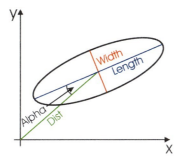

Figure 3.14: *Definition of the Hillas Parameters: the parameter Dist equals the geometric distance between the center of gravity of the Cherenkov light distribution and the source position in the camera. The angle between the major shower axis and the line joining the source location in the camera and the shower center of gravity is Alpha. The square root of the second moments of the light distribution along the major and minor axis of the ellipse are the parameters Width and Length.*

*Length, Dist, Alpha.*
The total amount of Cherenkov photons in the camera after the image cleaning, measured in equivalent photoelectrons, is described by the parameters *Size*. The fraction of light contained by the $n$ brightest pixels with respect to the total *Size* is called *Concentration[n]*. The sum of the signal charges (after image cleaning) of the pixels located in the outermost ring of pixels of the camera with respect to the total shower *Size* is defined as *Leakage*.

The parameters *MeanX* and *MeanY* correspond to the center of gravity (first moment) of the Cherenkov light distribution in the camera, the parameter *Dist* equals the geometric distance between the center of gravity and the source location in the camera. The angle between the major shower axis and the line joining the source location in the camera and the shower center of gravity is *Alpha*.

The square root of the second moments of the light distribution along the major and minor axis of the ellipse are the parameters *Width* and *Length*. The third moment of the light distribution along the major shower axis is *M3Long*.

### 3.2.5 Gamma/Hadron Separation

The MAGIC telescope does not only record $\gamma$-ray shower images, but it is also triggered by cosmic ray showers, single isolated muons and fluctuations from the light of the night sky. In fact, the background images are by a factor of up to several thousand more numerous than the images of $\gamma$-ray showers. Thus a statistical method has to be applied for the sample separation of $\gamma$-ray candidates (signal) and background events, see e.g. Fegan (1997) for a review.

As discussed in section 2.1.1, there are physical differences between hadronic and electromagnetic showers. In general, hadronic showers are broader, more irregular and subject to larger fluctuations. Moreover, the main axis of γ-ray shower images point to the source location in the camera, whereas the direction of the main axis for background showers is approximately uniformly distributed. Figure 3.15 shows two camera images, one of a MC simulated γ-ray shower and the other of a recorded background event. Figure 3.16 compares the image parameter distributions for MC γ-rays and OFF data (background) for Size > 200 photoelectrons. The distributions show substantial differences which provide a good γ/hadron separation power.

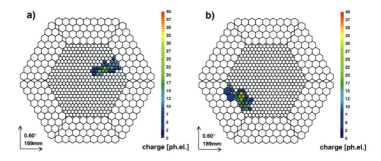

Figure 3.15: a) MC γ-ray shower image in the MAGIC camera, $E = 190$ GeV, Size = 410 photoelectrons. b) Background shower image, most probably due to a charged cosmic ray, Size = 410 photoelectrons.

In this analysis, a custom implementation (Bock et al. 2004; Hengstebeck 2003) of the Random Forest (RF) method (Breiman 2001) was applied for the γ/hadron separation. In the RF method to each event several not completely independent decision trees (series of cuts) are applied. By combining the results of the individual decision trees, the parameter hadronness is calculated, which is a measure of the probability that the event is not γ-ray like.

The trees of the RF are generated by means of training samples for the different classes: A sample of Monte Carlo (MC) generated γ-ray showers was used to represent the signal events together with randomly selected events drawn from the measured OFF-data to represent the background.

The source-position independent image parameters Size, Width, Length and Concentration are selected to parameterize the shower images. In addition, for sources at known locations the source-position dependent parameters Dist and M3Long are used. In order not to bias the RF towards a certain sky position, for the determination of unknown sources as well as extended sources the parameter Disp, a function of Size, Width and Length (see section 3.2.7) is used instead. Note that the Alpha parameter is not included in the RF training. In general, the Size distributions of MC generated γ-rays and measured OFF-data are different, which might influence the training of the Random Forest. Therefore,

## 3.2 Event Reconstruction

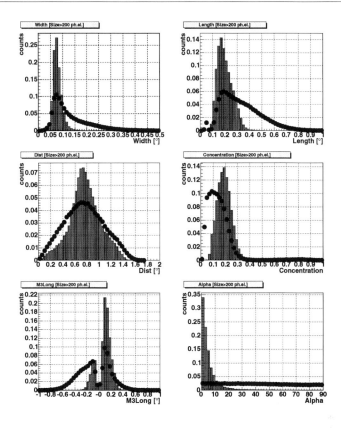

Figure 3.16: *Comparison of the distributions of the image parameter Width, Length, Dist, Concentration, third moment along the long axis and Hadronness for MC $\gamma$-rays (red histograms) with a power law spectrum with slope of -2.6 and OFF data (blue points) for Size > 200 photoelectrons. The distributions show substantial differences which provide a good $\gamma$/hadron separation power.*

a subsample of the OFF-data is randomly chosen which has the same Size distribution as the MC generated $\gamma$-rays. Only showers with a minimum Size of 200 photoelectrons were considered in the training. The $\gamma$-ray sample is defined by selecting showers with a hadronness below a specified value. An independent sample of MC $\gamma$-ray showers was used to determine the efficiency of the cuts, which depends on the hadronness cut value chosen.

Figure 3.17 displays the so-called Gini-index for the different image parameters used in the RF training. The Gini-index is a measure for the relative discrimination power of the

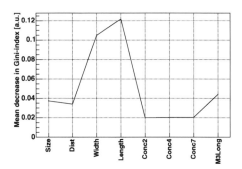

Figure 3.17: *The Gini index values for the RF training parameters are a measure for the relative importance of the parameters for the γ/hadron discrimination: The most important parameters are Width and Length. The Size parameter was used for the scaling of the other parameters but does not contribute itself to the γ-hadron separation, see text.*

different parameters. The most important parameters are Width and Length. The Size parameter was used for the scaling of the other parameters, but one has to make sure that it does not contribute itself to the γ-hadron separation. This is done by using MC γ-rays and measured background showers with the same Size distribution in the RF training.

Figure 3.18a shows the distribution of the hadronness parameter for MC simulated γ-ray showers and measured Crab Nebula ON and OFF-data (background). The hadronness for the simulated γ-rays peaks at small values, whereas the hadronness for background showers peaks at larger values. Nevertheless, there exist always a fraction of hadron induced showers which have an image nearly like a γ-ray shower (for example if in the first interaction nearly all energy is transferred to a $\pi^0$). For larger hadronness values the ON and OFF distributions agree well, for small hadronness values there is an excess of the ON data over the OFF data from the Crab Nebula γ-rays. Figure 3.18b shows the γ-ray cut efficiency as a function of the background efficiency, the so-called Neyman-Pearson plot. For a γ-ray efficiency of 50% only about 1.25% of the background events pass the cuts.

### 3.2.6 Energy Reconstruction

The total Cherenkov light intensity recorded with the camera of the MAGIC telescope for γ-ray showers depends on the energy of the primary γ-ray, the observation zenith angle, the impact parameter of the γ-ray with respect to the telescope axis and on the atmospheric extinction. For a given atmospheric extinction and observation zenith angle it is possible to estimate the primary γ-ray energy (and the impact parameter) from the shower image.

In the presented analysis a slightly modified Random Forest algorithm is used for the energy estimation. The image parameters Size, Width, Length, Dist, Concentration and Leakage as well as the observation zenith angle are used to characterize the shower images. Using MC simulated γ-ray showers (without applying γ/hadron separation cuts) the RF

## 3.2 Event Reconstruction

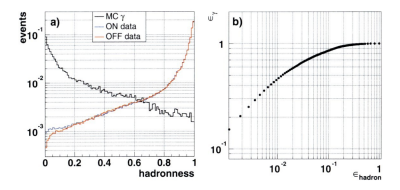

Figure 3.18: *a) Normalized hadronness distributions for MC $\gamma$-rays and measured Crab Nebula ON and OFF data: The MC $\gamma$-ray showers have a small hadronness, whereas the measured ON and OFF data peak for large hadronness values. For larger hadronness values the ON and OFF distributions agree well, for small hadronness values there is an excess of the ON data over the OFF data from the Crab Nebula $\gamma$-rays. b) Neyman-Pearson plot: the $\gamma$-ray efficiency of the Random Forest cuts as a function of the background efficiency. For a $\gamma$-ray efficiency of 50% only about 1.25% of the background events pass the cuts.*

is trained to separate the population of events with matching simulated energy from those having an energy outside a corresponding energy bin. The combination of image parameters determines the "probability" of an event to belong to a given energy bin and the one with the highest "probability" is selected.

Figure 3.19a displays the distribution of the simulated vs. RF reconstructed energies for MC $\gamma$-rays. For the lowest simulated energies there is a sizeable bias towards larger estimated energies. This is a consequence of the trigger which selects close above threshold only images with fluctuations towards a higher Size. All energy dependent distributions have to be corrected for the mis-match between estimated and true energy, see section 3.4.3. Figure 3.19b shows the distribution of the ratio between the difference of estimated and true energy over the true energy. For simulated energies above 100 GeV and Size > 200 photoelectrons the reconstructed energy is on average equal to the simulated one, the average energy resolution ($\sigma(E_{\text{reconstructed}} - E_{\text{simul.}})/E_{\text{simul.}}$) is about 25%.

### 3.2.7 Source Position Reconstruction: The Disp Method

Up to this point for each shower a measure for the probability to be a background event and an energy estimate (assuming it was a $\gamma$-ray shower) was calculated. A further task is the determination of the source position in the sky. This is especially important in case of uncertainties in the a priori knowledge of the source position (e.g. unidentified EGRET sources or GRBs), serendipitous searches for sources in the field of view (e.g. in a sky scan)

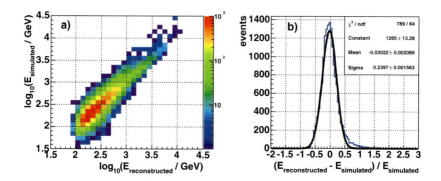

Figure 3.19: *RF energy estimation and energy resolution: a) Distribution of simulated vs. reconstructed energy for MC γ-rays. b) For simulated energies above 100 GeV and a minimum Size cut of 200 photoelectrons the average estimated (reconstructed) energy is equal to the MC simulated energy. The average energy resolution is about 25%, it decreases with increasing energy of the primary γ-ray.*

or the study of the morphology of extended sources. In order to produce a non-correlated map of the sky in VHE γ-rays, one has to assign to each event a unique source position in the sky (later called "arrival direction").

For each event the arrival direction of the primary in sky coordinates is estimated by using the Disp-method (Fomin et al. 1994; Lessard et al. 2001; Domingo-Santamaria et al. 2005): The arrival direction is assumed to lie on the major axis of the Hillas ellipse that fits the shower image in the camera at a certain distance (Disp) from the image center of gravity, see figure 3.20. Shower images which are closer to the source position in the camera are more roundish, whereas showers which are further away from the source position in the camera are more elongated. Thus Disp can be parameterized as a function of the ellipticity (Lessard et al. 2001):

$$\text{Disp} = \xi \cdot \left(1 - \frac{\text{Width}}{\text{Length}}\right), \quad \xi = A + B \cdot \log \text{Size} + C \cdot (\log \text{Size})^2 \;. \tag{3.11}$$

The parameters $A, B$ and $C$ are determined from MC simulations after application of γ/hadron separation cuts. A second order polynomial in log Size is fitted to the distribution of Disp/(1-Width/Length) as a function of log Size.

This Disp calculation provides two possible primary origins along the major shower axis. Therefore, the correct one has to be selected. Cherenkov photons from the upper part of the shower create a narrower section of the image with a higher photon density (head), photons from the lower part of the shower normally generate a much more fuzzy and spread image end (tail). Therefore, asymmetries in the charge distribution along the major axis of the images can indicate to which direction from the shower center of gravity the

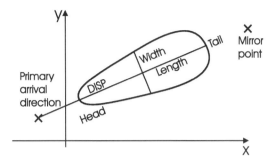

Figure 3.20: *The Disp method to reconstruct the primary arrival direction: The arrival direction is assumed to lie on the major axis of the Hillas ellipse that fits the shower image in the camera at a certain distance (Disp) from the image center of gravity. The asymmetry in the charge distribution along the major axis of the image indicates to which side from the shower center of gravity the source position is located.*

source position is located. In the presented analysis the third moment of the photoelectrons distribution along the major image axis, M3Long, is used. Note, that the precision of the head/tail determination using the parameter M3Long depends on the image Size. Above 500 photoelectrons the efficiency is well above 90%, below 200 photoelectrons the head/tail determination may be wrong for nearly 50% of the events. In general, the Disp method yields the arrival direction of the primary in camera coordinates. By knowing the telescope pointing direction and the observation time the camera coordinates can be converted to celestial coordinates (Wittek 2005b). A point source of $\gamma$-rays yields a reconstructed sky distribution of $\gamma$-rays which can be fitted by a two-dimensional Gaussian with a sigma of 0.1°, see section 3.6.2. In other words: The $\gamma$-ray PSF is about 0.1° when using the Disp method.

## 3.3 $\gamma$-Ray Signal Reconstruction / Background Subtraction

After calculating for each event the probability to be a background event (the so-called hadronness) a cut in the hadronness value is applied to separate $\gamma$-ray candidate events from background events. For the $\gamma$-ray candidate events the arrival direction of the primary and the energy are computed. The $\gamma$-ray signal and the background level of the observation have to be calculated for the observed data set using a statistical approach: In the observed data set there are $N_{ON}$ $\gamma$-ray candidate events with respect to the source; part of this number are genuine $\gamma$-rays from the source and the other part is due to various backgrounds producing $\gamma$-ray like air showers. The background is determined from the number of $\gamma$-ray candidates, $N_{OFF}$, with respect to an OFF-source, a sky region from where no $\gamma$-rays are expected.

The number of genuine $\gamma$-rays from the source, $N_\gamma$, is given as the excess of the ON-source $\gamma$-ray candidates over the scaled number of background $\gamma$-ray candidates:

$$N_\gamma = N_{\mathrm{ON}} - \alpha N_{\mathrm{OFF}} , \qquad (3.12)$$

where $\alpha$ is the normalization factor between the ON-source and the OFF-source data set. This method relies on the assumption that the systematic differences of the ON and OFF data sets are small compared to the $\gamma$-ray signal from the source. Moreover, the reconstructed $\gamma$-ray signal depends critically on the determination of the normalization factor $\alpha$.

The significance that the observed excess $\gamma$-ray signal from the source is not due to a background fluctuation is given by equation 17 of Li&Ma (1983)):

$$S_\gamma = \sqrt{2} \left( N_{\mathrm{ON}} \ln \left( \frac{(1+\alpha) N_{\mathrm{ON}}}{\alpha (N_{\mathrm{ON}} + N_{\mathrm{OFF}})} \right) + N_{\mathrm{OFF}} \ln \left( \frac{(1+\alpha) N_{\mathrm{OFF}}}{N_{\mathrm{ON}} + N_{\mathrm{OFF}}} \right) \right)^{1/2} . \qquad (3.13)$$

Note that this equation only considers statistical fluctuation of the ON and OFF $\gamma$-ray candidate counts and assumes an exact knowledge of the normalization constant $\alpha$.

In the following two analysis methods to determine the $\gamma$-ray signal ($N_{\mathrm{ON}}$, $N_{\mathrm{OFF}}$ and $\alpha$) are developed: the Alpha analysis (section 3.3.1) and the Disp sky map / $\theta^2$ analysis (section 3.3.2).

### 3.3.1 Alpha Analysis

One method to determine the normalization constant between the $\gamma$-ray candidates for the ON and OFF data set is by means of the distribution of the image orientation angle Alpha, see section 3.2.4. As shown in figure 3.16 $\gamma$-ray showers point to the source (they have low values of Alpha), whereas for background showers the Alpha parameter is in first order distributed uniformly between 0° and 90°. In case one knows the shower head/tail information, one can define Alpha to be from $-90°$ to 90°. For image sizes above 200 photoelectrons there is only a negligible number of $\gamma$-ray shower images expected with Alpha values above 30°. Thus the region Alpha > 30° contains only background events, both in the ON and OFF data, and can be used to determine the normalization constant $\alpha$. Figure 3.21 shows as an example the Alpha distributions for ON data of the Crab Nebula and dedicated OFF data for the background estimation. The two distributions have been normalized in the region 30° < Alpha < 80°. In this region the Alpha distributions of the ON and OFF data agree well with each other, no systematic differences can be seen. For Alpha < 7.5° there is an excess of 597 $\gamma$-ray candidate events of the ON data over the OFF data, corresponding to a significance of 18.8 $\sigma$. In case of using a fit to the OFF data distribution (e.g. with a second order polynomial) in the full Alpha range one can determine the background for Alpha < 7.5° with a higher statistical precision. Thereby one obtains a significance of 24.1$\sigma$.

The width of the Alpha distribution for $\gamma$-rays depends strongly on the image parameter Size. A fit to the alpha distribution with a Gaussian centered at zero degrees yields a sigma of around 2° for Size > 600 photoelectrons. For small Sizes below 150 photoelectrons the sigma value may be larger than 10°. To achieve the highest significance, the cut in the

## 3.3 γ-Ray Signal Reconstruction / Background Subtraction

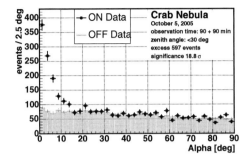

Figure 3.21: *Distributions of the image orientation angle Alpha for ON data of the Crab Nebula and dedicated OFF data for Size > 400 photoelectrons, corresponding to an analysis energy threshold of about 280 GeV. The distributions have been normalized to each other in the region $30° <$ Alpha $< 80°$. In this region the Alpha distributions of the ON and OFF data agree well, no systematic differences can be seen. For Alpha $< 7.5°$ there is an excess of 597 γ-ray candidate events of the ON data over the OFF data, corresponding to a significance of $18.8\sigma$.*

parameter Alpha may be adjusted to about 1.8 the sigma of the γ-ray Alpha distribution (Li&Ma 1983).

### 3.3.2 Disp-Sky Map Analysis

Using the Disp method for each event the arrival direction of the primary particle that initiates the air shower can be reconstructed in camera as well as in sky coordinates. In order not to bias the arrival direction calculation to any position, only source position independent image parameters are used in the RF training of the γ/hadron separation, see section 3.2.5. For example, the Dist parameter cannot be used.

Figure 3.22a) and b) show the reconstructed arrival direction for γ-ray candidate events (after γ/hadron separation cuts) in camera and sky coordinates, respectively, for ON-source data of the Crab Nebula taken on October 5, 2005. A minimum image Size of 200 photoelectrons was required. In addition to reconstructed γ-rays from the source, there are background events, which fulfill the selection criteria, i.e. the cut in the hadronness parameter.

This section is structured as follows. In section 3.3.2.1 the background distribution in the sky map is calculated according to the camera acceptance. In section 3.3.2.2 the source position is determined from the background subtracted sky map and in section 3.3.2.3 the squared angular distributions ($\theta^2$ distributions) between the reconstructed shower direction and the source are studied for ON data and background. Thereafter the software pointing correction using the MAGIC starguider is explained in section 3.3.2.4. Finally, in section 3.3.2.5, possibilities for the smoothing of the reconstructed sky maps are discussed.

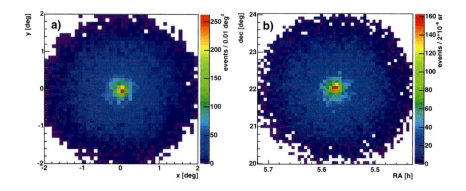

Figure 3.22: *Raw Disp maps (after γ/hadron separation and without background subtraction) a) in camera and b) in sky coordinates, the Crab Nebula γ-ray source is located in the center of the camera. A minimum image Size of 200 photoelectrons was required.*

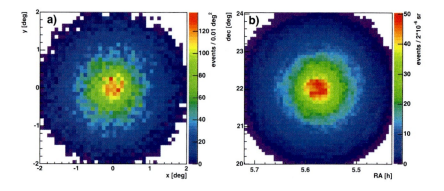

Figure 3.23: *a) Disp map in camera coordinates without a known γ-ray source in the field of view (OFF data). The data reflect the camera acceptance for a uniform background of γ-ray candidates. b) Generated background sky map corresponding to the ON-data of figure 3.22. Note, that for the chosen Size cut of 200 photoelectrons the camera acceptance is quite uniform in azimuth.*

### 3.3.2.1 Camera Acceptance / Background Determination

In general, the background in the sky map of the ON data depends on the intrinsic camera acceptance as well as on the particular star field of the ON region and observation conditions

## 3.3 γ-Ray Signal Reconstruction / Background Subtraction

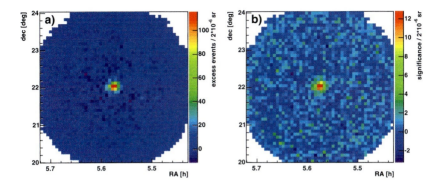

Figure 3.24: *a) Background subtracted sky map of the Crab Nebula, the color scale shows the reconstructed brightness distribution in VHE γ-rays. A minimum image Size of 200 photoelectrons was required. b) The same data shown as significance per sky bin, the sky bins are statistically independent.*

(especially observation ZA and weather conditions). In the following it is assumed that the OFF-source data match the sky brightness and observation conditions well. Only the intrinsic camera acceptance is considered for the background determination.

In order to determine the background in the ON data, a Disp sky map in camera coordinates is computed with the same algorithm but for a data set without a known γ-ray source in the field of view. Figure 3.23a depicts the Disp map in camera coordinates for an observation without γ-ray source in the field of view, thus this map just shows the camera acceptance to a uniform background of γ-ray candidates. In this particular case the observation of an AGN candidate source on October 5, 2005, where no signal was found, was used as OFF data as no dedicated OFF data in the Galactic plane has been taken. For a discussion about the derivation of the best suited background model see section 3.5.

To determine the background in a certain ON-source data set, for each ON-source event the camera coordinates of a background event are generated taking the above determined camera acceptance map as probability density function for the background events in the camera. Thereafter, the camera coordinates of the generated background event are projected into sky coordinates using the same telescope pointing direction and time as for the ON-source event. By construction, the generated background events have the same time distribution as the ON-data and the same distribution of estimated source positions in the camera as the OFF data. Figure 3.23b shows the obtained background sky map corresponding to the ON-source observation in figure 3.22.

The normalization factor $\alpha$ between the ON-source sky map and the background sky map can be computed as the ratio between the total number of γ-ray candidate event counts in the ON-source sky map with a minimum angular distance to the source location and the corresponding number of entries in the background sky map. Figure 3.24a displays the background subtracted sky map of the Crab Nebula in units of excess events per solid angle, while figure 3.24b shows the same sky region in significance per solid angle bin. A $\gamma$-ray excess can clearly be seen at the position of the Crab Nebula (RA = $5^\mathrm{h}34^\mathrm{m}31^\mathrm{s}$, dec= $22°0'52''$), whereas the rest of the sky map is consistent with no $\gamma$-ray signal.

#### 3.3.2.2 Source Position Reconstruction

In order to determine the source position ($\mathrm{RA_{source}}, \mathrm{dec_{source}}$) and extension ($\sigma_\mathrm{source}$), the background subtracted sky map (VHE $\gamma$-ray sky brightness map $b(\mathrm{RA}, \mathrm{dec})$) is fitted by the following two-dimensional Gaussian:

$$b(\mathrm{RA}, \mathrm{dec}) = C \exp\left[-\frac{(\mathrm{RA} - \mathrm{RA_{source}})^2 \cdot \cos^2(\mathrm{dec}) + (\mathrm{dec} - \mathrm{dec_{source}})^2}{2(\sigma_\mathrm{source}^2 + \sigma_\mathrm{PSF}^2)}\right] . \quad (3.14)$$

Such a fit to the Crab Nebula sky map presented in figure 3.24a results in:

$$\mathrm{RA_{Crab}} = 5^\mathrm{h}34^\mathrm{m}30^\mathrm{s} \pm 2^\mathrm{s}, \quad \mathrm{dec_{Crab}} = 20°2'24'' \pm 23'' \text{ and } \sqrt{\sigma_\mathrm{Crab}^2 + \sigma_\mathrm{PSF}^2} = 0.95° \pm 0.06° . \quad (3.15)$$

Within the systematic telescope pointing uncertainty of 2', see section 2.2.2, the reconstructed position is compatible with the nominal position of the Crab Nebula: RA = $5^\mathrm{h}34^\mathrm{m}31^\mathrm{s}$, dec= $22°0'52''$ (Han & Tian 1999). The VHE $\gamma$-ray emission is compatible with a point source emission for $\sigma_\mathrm{PSF} = 0.1°$. The angular resolution of the MAGIC telescope is discussed in section 3.6.2.

#### 3.3.2.3 $\theta^2$ Distributions

In section 3.3.1 the so-called Alpha analysis was presented, which allows to determine the normalization constant between the $\gamma$-ray candidates for the ON and OFF data set. Another method to determine this normalization constant is to compare the radial distributions of the reconstructed arrival directions for the ON and OFF data with respect to the candidate source:

Let $\theta$ be the angular distance between the reconstructed primary $\gamma$-ray candidate sky position and the known sky position of the $\gamma$-ray source. Figure 3.25 shows the distributions of $\theta^2$ for the ON-data (corresponding to figure 3.22) and the OFF data (corresponding to figure 3.23a) as well as the generated background model (corresponding to figure 3.23b. The $\theta^2$ distributions of the OFF data and the one generated from the background model agree well with each other. The ON and OFF data have been normalized for $0.2 \text{ deg}^2 \leq \theta^2 \leq 0.5 \text{ deg}^2$. In this region the distribution of the ON data agrees well with the ones of the OFF data and the background model, no systematic differences can be seen. For $\theta^2 < 0.04 \text{ deg}^2$ there is an excess of 552 $\gamma$-ray candidate events of the ON data over the OFF data, corresponding to a significance of $17.6\,\sigma$. In case of using a fit to the OFF data distribution (e.g. with a second order polynomial) in the full $\theta^2$ range one can determine the background for $\theta^2 < 0.04 \text{ deg}^2$ with a higher statistical precision. Thereby one obtains a significance of $22.0\sigma$.

## 3.3 γ-Ray Signal Reconstruction / Background Subtraction

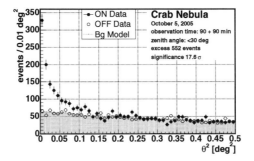

Figure 3.25: *Distributions of $\theta^2$ for the ON-source observation and dedicated OFF data (corresponding to the data set of figure 3.21) as well as the generated background model for Size > 400 photoelectrons, corresponding to an analysis energy threshold of about 280 GeV. In the normalization region $0.2\ \mathrm{deg}^2 \leq \theta^2 \leq 0.5\ \mathrm{deg}^2$ all distributions agree well, no systematic differences can be seen. For $\theta^2 < 0.04\,\mathrm{deg}^2$ there is an excess of 552 γ-ray candidate events of the ON data over the OFF data, corresponding to a significance of $17.6\,\sigma$.*

Both the Alpha analysis, see section 3.3.1, as well as the $\theta^2$ analysis can be used to determine the number of γ-ray candidate excess events of a data set. For the Alpha analysis one usually makes use of the known source position by including source dependent parameters (like Dist) in the Random Forest for the γ/hadron separation. For the Disp/$\theta^2$ analysis one usually does not use source position dependent parameters in the Random Forest training. The better γ/hadron separation power of the Random Forest with source-dependent parameters compared to the one without source-dependent parameters is compensated by the usage of a two-dimensional information of the source location per shower compared to the only one-dimensional information in the Alpha analysis. In conclusion, the Alpha and Disp/$\theta^2$ analysis give comparable sensitivities (significance per observation time).

#### 3.3.2.4 Software Pointing Correction Using the MAGIC Starguider

In order to perform morphological studies of celestial objects emitting VHE γ-rays the actual pointing direction of the MAGIC telescope must be monitored with a high precision compared to the γ-ray PSF of about 0.1°. Using the MAGIC star field monitor (see section 2.2.2), the actual pointing direction is determined every 10 some seconds by comparing the observed star field with the star catalogue position (Riegel et al. 2005). For the data presented in this thesis there was no absolute calibration of the pointing direction of the MAGIC star field monitor with respect to the pointing direction of the MAGIC telescope yet.

Figure 3.26 shows the pointing deviation between the intended telescope pointing position and the actual position for a 90 min observation of the Crab Nebula under low zenith

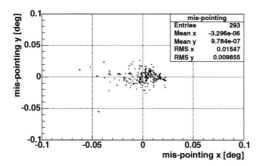

Figure 3.26: *Pointing Deviation as determined with the MAGIC starguider for the Crab Nebula data set of figures 3.21 and 3.25. For most observations the mis-pointing is small compared to the MAGIC $\gamma$-ray PSF of 0.1°. Nevertheless, possible mis-pointings (e.g. during source culmination) can efficiently be monitored and corrected for.*

angles. The mis-pointing is small compared to the MAGIC $\gamma$-ray PSF of 0.1°. Nevertheless, during source culmination there may be sizeable mis-pointings of about 0.1°.

### 3.3.2.5 Sky Map Folding and Interpolation Procedure

In figure 3.24 relatively coarse sky bins of $0.05° \times 0.05°$ have been chosen such that the source $\gamma$-rays are only spread out to a few bins (the MAGIC $\gamma$-ray PSF is about 0.1°) and the excess per bin is significant also for fainter sources than the Crab Nebula. Nevertheless, it is desirable to present the $\gamma$-ray sky images without this coarse binning. There are three possibilities:

- finer sky binning
- interpolation of the coarsely binned sky map
- folding of the coarsely binned sky map.

The first possibility requires increased statistics such that also for finer bins the signal in each sky bin is significant. Figure 3.27 illustrates the two latter possibilities to present the background subtracted sky map in a smoother fashion: In the interpolation case the $\gamma$-ray sky brightness $b(x, y)$ in the sky direction $(x, y)$ with $x_i \leq x < x_{i+1}$ and $y_i \leq y < y_{i+1}$ is approximated to be:

$$b(x,y) = \frac{(x_2 - x)(y_2 - y)b(x_1, y_1) + (x_2 - x)(y - y_1)b(x_1, y_2) + (x - x_1)(y_2 - y)b(x_2, y_1) + (x - x_1)(y - y_1)b(x_2, y_2)}{(x_2 - x_1)(y_2 - y_1)} \quad . \tag{3.16}$$

Here $(x_i, y_j)$ is the center of bin $(i, j)$ and $b(x_i, y_i)$ is the $\gamma$-ray sky brightness in bin $(i, j)$.

## 3.4 Determination of the γ-Ray Energy Spectrum

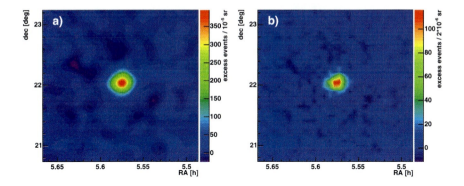

Figure 3.27: *Background subtracted γ-ray sky maps of the Crab Nebula: a) sky map of figure 3.24 folded with a two-dimensional Gaussian with σ = 0.072°, b) the bin centers of figure 3.24 interpolated.*

The folding of the sky map serves to increase the signal-to-noise ratio by effectively averaging the sky brightness over a larger sky area and thus reducing the statistical fluctuations. However, it degrades the spatial resolution. For presentation purposes the coarsely binned sky maps are folded with a two-dimensional Gaussian with a standard deviation of $\sigma_{2D} = 0.072°$ and a maximum of one:

$$b_{\text{folded}}(x,y) = \int_{-\infty}^{\infty} dX \int_{-\infty}^{\infty} dY \, b_{\text{binned}}(X,Y) \exp\left[-\frac{(X-x)^2 + (Y-y)^2}{2\sigma_{2D}}\right], \quad (3.17)$$

where $b_{\text{binned}}(X,Y)$ is the average γ-ray sky brightness in the sky bin containing the sky direction $(X,Y)$. For presentation of sky maps with low statistics the folding method is preferred over the interpolation method.

## 3.4 Determination of the γ-Ray Energy Spectrum

One of the most important results of the γ-ray observations is the energy spectrum $\Phi(E)$ of the VHE γ-ray sources. It provides valuable information on the acceleration processes of VHE particles in the astrophysical object. The differential energy spectrum (or differential γ-ray flux) of a source is defined as the number of γ-rays detected from the source, $N_\gamma$, per active detection area, per time and per energy interval:

$$\Phi(E) = \frac{dN_\gamma}{dE \, dA_{\text{eff}}(E) \, dt}. \quad (3.18)$$

In section 3.4.1 the effective detection area is calculated from MC simulations and in section 3.4.2 the effective observation time is calculated. Finally in section 3.4.3 the effect of the instrumental energy resolution is corrected for by an unfolding algorithm and the source energy spectrum is calculated.

### 3.4.1 Effective Collection Area

The effective collection area $A_{\text{eff}}(E)$ describes the hypothetical area within which the MAGIC telescope would observe each $\gamma$-ray entering this area (100% efficiency). In practice it is the $\gamma$-ray reconstruction efficiency integrated over the plane perpendicular to the telescope axis:

$$A_{\text{eff}}(E, \phi, \theta_{ZA}) = \int_0^{2\pi} \int_0^{\infty} \epsilon(E, \phi, \theta_{ZA}, b) b \, db \, d\phi \,, \qquad (3.19)$$

where $E$ is the energy of the primary $\gamma$-ray and $\epsilon$ is the $\gamma$-ray reconstruction efficiency, which is a function of the energy, the zenith angle of the observation $\theta_{ZA}$, the azimuth angle $\phi$ and the impact parameter $b$.

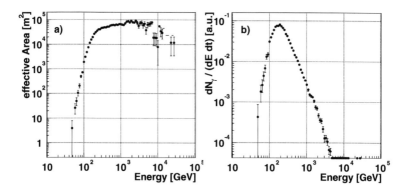

Figure 3.28: *a) Effective collection area for $\gamma$-rays as a function of the simulated $\gamma$-ray energy for zenith angles between 0° and 30° and Size > 200 photoelectrons. The source is located in the camera center. b) Corresponding expected differential $\gamma$-ray rate from a source with a spectrum of $E^{-2.6}$ as a function of the simulated $\gamma$-ray energy.*

The effective Collection Area is computed from MC simulations, as

$$A_{\text{eff}}(E, \phi, \theta_{ZA}) = \frac{N_\gamma^{\text{reconstructed}}(E, \phi, \theta_{ZA})}{N_\gamma^{\text{simulated}}(E, \phi, \theta_{ZA})} \times A_{\text{simulated}} \,, \qquad (3.20)$$

where it is assumed that a large enough area ($A_{\text{simulated}}$) was simulated such that beyond this area no $\gamma$-rays are reconstructed (the $\gamma$-ray reconstruction efficiency is equal to zero).

Figure 3.28a displays the effective collection area as a function of the simulated MC $\gamma$-ray energy for ZAs between 0° and 30° and with a lower Size cut of 200 photoelectrons,

## 3.4 Determination of the γ-Ray Energy Spectrum

the source is in the center on the camera. Figure 3.28b shows the corresponding expected observed differential γ-ray rate from a source with a spectrum proportional to $E^{-2.6}$ as a function of the simulated γ-ray energy. At low energies the effective area and the expected differential γ-ray rate show a sharp rise with energy. For these lowest energies the effective area is limited by the amount of Cherenkov light collected by the telescope to pass the Size cut. The energy where the expected differential γ-ray rate reaches its maximum is defined as the energy threshold of the telescope, see also section 3.6.3. For larger energies there is a slow variation of the effective collection area with energy. Only for the largest energies the effective area drops due to leakage of the shower pictures out of the MAGIC camera. At present there are dedicated efforts to lower the analysis energy threshold of the MAGIC telescope. As the analysis presented in this thesis does not critically depend on a lowest possible energy threshold a conservative lower Size cut of 200 photoelectrons has been chosen.

### 3.4.2 Effective Observation Time

The effective observation time is defined as the time within which the observed number of events would have been observed with an ideal detector. This is equivalent to the time during which the telescope was ready to record an event. Due to some small dead-times when the DAQ system cannot accept a trigger, the effective time when the DAQ system can record an air shower is slightly less than the time difference between the run start and stop time. As the dead-time is not a constant, the effective observation time is determined by a fit (Wittek 2002a) to the distribution of the observed time differences $t$ between consecutive recorded air showers (Poisson statistics):

$$\frac{\Delta N_{\text{recorded}}}{\Delta t} \propto \exp\left[-R\,t\right], \tag{3.21}$$

where $\Delta N_{\text{recorded}}$ is the number of recorded shower arrival time differences in the time interval $[t, t + \Delta t]$ and $R$ is the average shower rate assuming a vanishing dead time. In order to yield the true air shower rate, the fit has to be performed only for time differences much larger than the maximum dead time. On the other hand, there may also be a few very large time differences (in the order of seconds), e.g. due to the start of a new run, which must not be taken into account in the fit. Errors may arise when the air shower rate varies within the observation time (Wittek 2002a). Thus the effective observation time $t_{\text{eff}}$ is given by:

$$t_{\text{eff}} = N_{\text{recorded}}/R . \tag{3.22}$$

Figure 3.29 presents the distribution of the time differences between two consecutive recorded air shower events with Size > 200 photoelectrons. In total 179572 events are recorded. For large time differences the rate is determined to be $32.76 \pm 0.12$ Hz, which leads to an effective observation time of $5410 \pm 19$ s. This has to be compared to the cumulated time differences between run starts and stops of 5508 s, which yields a dead-time in the order of 2% of the total observation time.

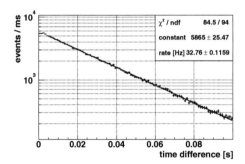

Figure 3.29: *Effective observation time: Distribution of time differences between two consecutive recorded air showers with an image Size > 200 photoelectrons. The distribution can be fitted well for time differences above a few μs by equation (3.21). Smaller time differences are affected by dead-times. A rate of* $32.76 \pm 0.12$ *Hz and a total of 179572 recorded events yield an effective observation time of* $5410 \pm 19$ *s, compared to the cumulated time differences between run stops and run starts of 5508 s.*

### 3.4.3 Unfolding of the Energy Spectrum

With the MAGIC telescope any measurement can only be done as a function of the estimated energy of the primary $\gamma$-ray, the most important measurement being the number of $\gamma$-ray excess events as a function of estimated energy. Due to experimental deficiencies, the experimentally measured (estimated) value of the energy, $E_{\text{estimated}}$ is not identical with the true one, $E_{\text{true}}$. As a consequence, the measured distribution in $E_{\text{estimated}}$ is a convolution of the true distribution with a resolution function describing the deficiencies of the experiment. The aim of the unfolding is to recover the true distribution in $E_{\text{true}}$ from the measured one, using a resolution function which has been determined from MC simulations. For further details about the unfolding algorithm see Wittek (2005a); Anykeev et al. (1991) and references therein. The implemented unfolding routines in the MAGIC software framework MARS are described by Aliu & Wittek (2006).

As an example figure 3.30a shows the distribution of $\gamma$-ray excess events from the direction of the Crab Nebula as a function of the estimated energy. Figure 3.30c shows the result of the unfolding, the distribution of $\gamma$-ray excess events as a function of true energy. The bins in true energy are chosen larger than in estimated energy to reduce the number of unknowns in the unfolding procedure. Figure 3.30b illustrates the energy resolution function, the distribution of simulated vs. reconstructed energy for MC $\gamma$-rays after all data analysis cuts. The blue lines indicate the bins in estimated energy, which were used in the unfolding algorithm. In this case 20 bins with estimated energies between 90 GeV and about 1700 GeV are chosen. Measured excess events with higher energies are not taken into account due to low MC statistics at large energies. There are no excess events with a smaller reconstructed energy. The red lines indicate the 11 reconstructed bins in true

## 3.4 Determination of the γ-Ray Energy Spectrum

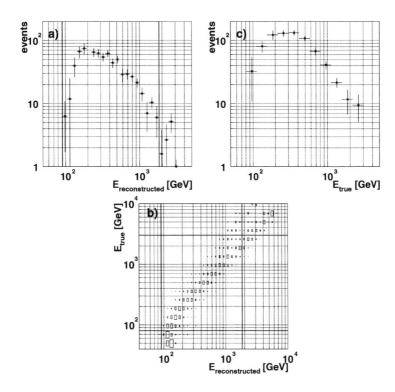

Figure 3.30: *Distribution of γ-ray excess events from the Crab Nebula as functions of a) the estimated energy and c) the true energy. b) Distribution of simulated vs. reconstructed energy for MC γ-rays after all data analysis cuts. The blue lines indicate the bins in estimated energy, which were used in the unfolding algorithm. The red lines indicate the reconstructed bins in true energy, see text.*

energy (energies between 80 GeV and 3000 GeV). The number of reconstructed bins in true energy is well below the number of measured bins in estimated energy. Although the reconstructed range in true energy is larger than the chosen range in estimated energy, there is a non-zero energy migration from every of the true energies into the chosen range of estimated energies.

Figure 3.31 displays the reconstructed very high energy γ-ray spectrum $(dN_\gamma/(dE_\gamma dAdt))$ vs. true $E_\gamma$) of the Crab Nebula after correcting (unfolding) for the instrumental energy resolution. The horizontal bars indicate the bin size in energy, the marker is placed in the bin center on a logarithmic scale.

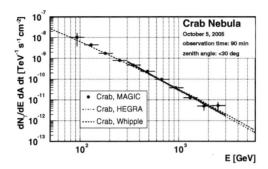

Figure 3.31: *Energy spectrum of the Crab Nebula. Within errors it agrees with the previous measurements by the HEGRA collaboration (Aharonian et al. 2004a) (E > 500GeV) and the spectral model fit to the Whipple data (E > 300 GeV) by Hillas et al. (1998). The full line shows a fit to the MAGIC data taking the full instrumental energy resolution into account, see text.*

Above an energy of about 300 GeV the energy spectrum of the Crab Nebula can be approximated by a simple power law (Hillas et al. 1998; Aharonian et al. 2004a): The full line shows the result of a forward unfolding procedure: A simple power law spectrum is assumed for the true differential $\gamma$-ray flux. The parameters of the power law are determined by fitting the predicted differential flux to the measured energy spectrum $(dN_\gamma/(dE_\gamma dAdt)$ vs. estimated $E_\gamma$) taking the full instrumental energy migration (true $E_\gamma$ vs. estimated $E_\gamma$) into account as described by Wittek (2005a) and Mizobuchi et al. (2005). The result of the fit in the region $300\,\text{GeV} < E_{\text{estimated}} < 2\,\text{TeV}$ is given by ($\chi^2/\text{n.d.f} = 13.5/10$):

$$\frac{dN_\gamma}{dAdtdE} = (3.0 \pm 0.3) \times 10^{-11} \left(\frac{E}{\text{TeV}}\right)^{-2.6\pm0.1} \text{cm}^{-2}\text{s}^{-1}\text{TeV}^{-1}.$$

The given errors ($1\sigma$) are purely statistical. The systematic error is estimated to be 35% in the integral flux level and 0.2 in the spectral index, see section 3.7. The measured energy spectrum of the Crab Nebula agrees well within errors with the previous measurements by the HEGRA collaboration (Aharonian et al. 2004a) ($E > 500\text{GeV}$) and the spectral model fit to the Whipple data ($E > 300$ GeV) (Hillas et al. 1998), as well as to an independent analysis of the MAGIC data (Wagner et al. 2005). As the Crab Nebula is a constant source of $\gamma$-rays, this result gives confidence in the analysis methods. The analysis has not been optimized to reconstruct the low energy part of the spectrum.

## 3.5 Analysis of Data Taken in the Wobble Mode

As explained in section 2.3, there are two modes of data taking with the MAGIC telescope: ON/OFF observations and the so-called wobble mode. Thereby the telescope tracks a sky

## 3.5 Analysis of Data Taken in the Wobble Mode

position close to the candidate source, typically at an offset of 0.4°, see e.g. Bretz et al. (2005). The source rotates around the camera center due to the Altitude-Azimuth telescope mount. This requires a dedicated analysis procedure, which was developed in this thesis and will be presented below:

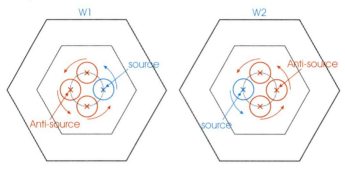

Figure 3.32: *Wobble mode observation: The source and anti-source position are located opposite with respect to the camera center. Due to the Altitude-Azimuth mount of the telescope source and anti-source rotate in the camera. The blue region around the source defines the ON-region, whereas all three red regions can be used as background control regions. When changing the wobble position W1 to W2 the source and anti-source positions in the camera are exchanged.*

Figure 3.32 illustrates the location of the source position in the camera for the two wobble tracking positions W1 and W2. The mirrored source position with respect to the camera center is called "anti-source" position. The possible off-source tracking positions (same offset to the candidate source) are switched regularly in order to make the distributions of the source and anti-source positions in the camera as similar as possible. Thus systematic effects due to inhomogeneities of the camera acceptance cancel out for parameter distributions with respect to the source and anti-source. Nevertheless, the switching from the tracking position W1 to W2 does not mirror-image the source on top of the anti-source and vice-versa in the camera. The switching from W1 to W2 is rather a translational displacement of the sky in the camera. The sky positions projected to the anti-source positions in the camera are not the same for W1 and W2. Therefore, the wobble mode only cancels out inhomogeneities in the camera, but not inhomogeneities of the sky (e.g. bright stars).

One can define two more "background control" position (see e.g. Rowell (2003)) by a 90° rotation of source and anti-source around the camera center. Assuming a rotational symmetry of the camera acceptance around the camera center, one can use the anti-source position as well as the additional two background control positions to evaluate the background at the source position. Thereby one can measure ON and OFF data at the same time. This allows a reliable background estimation independent of possible changes in observation conditions like the weather.

This section is structured as follows: in section 3.5.1 the effect of the relatively small MAGIC trigger region on the wobble observations is discussed. Thereafter, in section 3.5.2 the ON and OFF samples are defined from the wobble data and the γ-ray signal is determined. Finally in section 3.5.3, the ON/OFF and the wobble observation modi are compared.

## 3.5.1 Trigger Losses in Effective Area

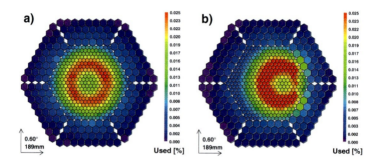

Figure 3.33: *Charge signal distribution in the MAGIC camera for MC simulated γ-rays, with a source a) located in the camera center and b) located 0.4° off-set from the center in the positive x direction. While figure a) shows a rotational symmetric charge distribution, in figure b) there is an apparent asymmetry due to the trigger losses for shower image center of gravities outside the trigger region of the MAGIC camera.*

For wobble mode observations, as well as for observations of extended sources, the finite trigger region of about 0.9° radius of the MAGIC telescope, see section 2.2.5, is of major concern. Figure 3.33 presents the charge signal distribution in the MAGIC camera for MC simulated γ-rays, with a source a) located in the camera center and b) located 0.4° off-set from the center in the positive x direction. While figure a) shows a rotational symmetric charge distribution, in figure b) there is an apparent asymmetry due to the trigger losses for shower image center of gravities outside the trigger region of the MAGIC camera.

The trigger loss is energy dependent. Higher energy showers have a higher chance to extend inside the trigger region of the MAGIC camera. Figure 3.34 depicts the ratio of the effective collection area after the image cleaning between wobble mode observations (0.4° source off-set) and observations with the source in the camera center as a function of the simulated γ-ray energy. The ratio rises with energy, above an energy of 200 GeV the effective area in the wobble mode is reduced by less than 10% compared to ON observation.

The losses in the effective area can be circumvented by enlarging the trigger region of the camera. This is one of the design considerations for the camera of the second MAGIC telescope (Teshima et al. 2005).

## 3.5 Analysis of Data Taken in the Wobble Mode

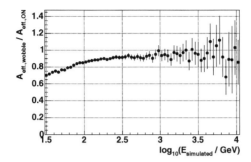

Figure 3.34: *Ratio of the trigger efficiency (effective area) of wobble mode and ON observations, from MC simulations. The ratio rises with energy. Above an energy of 200 GeV the effective area in the wobble mode is reduced by less than 10% compared to ON observation.*

### 3.5.2 Definition of ON and OFF samples for the Wobble Mode, Gamma Signal Determination

In the wobble mode observations, both the source and background control regions are observed simultaneously at camera positions where the acceptance is assumed to be equal. The $\gamma$-ray signal is obtained as excess of the $\gamma$-ray candidates from the source direction (ON data) over the background (OFF data). Therefore, a statistically independent definition of the ON and OFF data samples is necessary, either using the Alpha analysis (section 3.5.2.1) or using the $\theta^2$ analysis (section 3.5.2.2).

#### 3.5.2.1 Alpha Analysis

One possibility to determine the $\gamma$-ray signal is to use the image orientation angle Alpha with respect to the source and anti-source position. Figure 3.35 shows the definition of the Alpha and anti-Alpha angle. For each event the parameter Alpha is calculated for both positions. The $\gamma$-ray signal is expected to be an excess for low values of the Alpha distribution over the anti-Alpha distribution. However, both distributions are not statistically independent as one particular shower may have a small Alpha value as well as a small anti-Alpha value. In order to produce statistically independent ON and OFF samples and to suppress any signal contribution to the OFF sample, a cut on the anti-Alpha parameter is applied: An event is only included into the ON distribution in case the Alpha value with respect to the anti-source is above a certain limit, which is generally taken to be 15°. A corresponding cut is applied to the OFF distribution. This anti-Alpha cut causes an additional loss of effective area in the wobble mode observations (for the trigger losses see section 3.5.1 above) compared to ON/OFF observations, where no anti-Alpha cut is necessary.

Figure 3.36 shows as an example the distributions for the image orientation angle Alpha

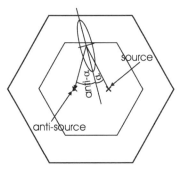

Figure 3.35: *Definition of the image orientation angles Alpha and anti-Alpha for wobble data: Alpha is the angle between the line joining the source position in the camera and the center of gravity of the shower image and the long axis of the shower ellipse. Anti-Alpha is the corresponding angle with respect to the Anti-source position in the camera (source position mirrored at the camera center).*

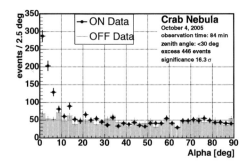

Figure 3.36: *Distributions for the image orientation angle Alpha with respect to the source (ON Data) and to the anti-source (OFF Data) for Crab Nebula wobble observations. A lower Size cut of 400 photoelectrons has been applied (analysis energy threshold about 280 GeV). An upper anti-Alpha cut of* 15° *has been applied, such that the ON and OFF distributions are independent for Alpha* < 15° *and correlated for Alpha values above. No normalization between ON and OFF has been applied, for Alpha* > 30° *the ON and OFF distributions agree well with each other. For Alpha* < 7.5° *there is an excess of 446 γ-ray candidate events of the ON over the OFF data, corresponding to a significance of* 16.3 σ.

## 3.5 Analysis of Data Taken in the Wobble Mode

with respect to the source (ON Data) and to the anti-source (OFF Data) for Crab Nebula wobble observations. No normalization between the ON and OFF data has been applied. For Alpha $< 7.5°$ there is an excess of 446 $\gamma$-ray candidate events of the ON over the OFF data, corresponding to a significance of $16.3\,\sigma$. In case of using a fit to the OFF data distribution (e.g. with a second order polynomial) one obtains a significance of $20.9\,\sigma$.

### 3.5.2.2 Disp / $\theta^2$ analysis

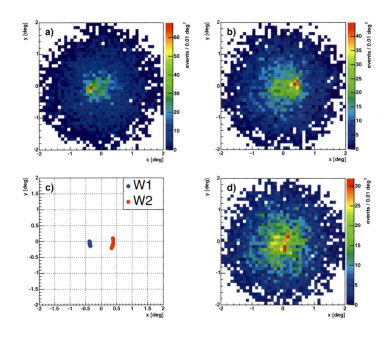

Figure 3.37: *a),b) Raw Disp maps (after $\gamma$/hadron separation and without background subtraction) in camera coordinates for the W1 and W2 observations of the Crab Nebula. c) The position of the Crab Nebula in the camera for both wobble observations. d) The background acceptance of the camera, derived from the background region of the two wobble observation positions, see text.*

Alternatively to the Alpha analysis, the $\gamma$-ray signal can also be evaluated using the Disp/$\theta^2$ analysis (see section 3.3.2) for the wobble data. Figures 3.37a) and b) show raw Disp maps (after $\gamma$/hadron separation and without background subtraction) in camera

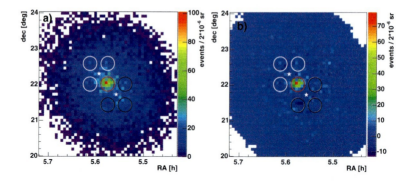

Figure 3.38: *a) Raw Disp map (after γ/hadron separation and without background subtraction) in sky coordinates for the combined W1 and W2 data set of the Crab Nebula. The two white stars show the two pointing positions for W1 and W2 observations. The red circle shows the source integration region. The white circles show the background control regions (same camera acceptance as the source region) for W1 and the black circles show the background control regions of W2. b) The same data set after background subtraction.*

coordinates for the W1 and W2 observations of the Crab Nebula, while figure 3.37c shows the location of the Crab Nebula in the Camera. One can see an excess of γ-ray candidate events in both figures 3.37a) and b) at the location of the Crab Nebula. For W1 (W2) observations there are no γ-rays from the Crab Nebula in the camera part with $x > 0$ ($x < 0$). Scaling these two camera parts to a common number of events and adding them, one obtains the background acceptance map of the camera, displayed in figure 3.37d).

Figure 3.38a exhibits a raw Disp map (after γ/hadron separation and without background subtraction) in sky coordinates for the combined W1 and W2 data set of the Crab Nebula. The two white stars show the two pointing positions for W1 and W2 observations. One can see an excess of γ-ray candidate events at the position of the Crab Nebula (RA = $5^h34^m31^s$, dec= $22°0'52''$) over the background. The red circle (radius 0.1° corresponding the the MAGIC PSF) shows the source integration region. For a telescope pointing to W1, the white circles are located rotationally symmetric in the camera to the source region. Assuming a rotationally symmetric camera acceptance, the events in the white circles can be used to estimate the background at the source position. They are called background control regions. Similarly for a telescope pointing to W2 the three black circles can be used as background control regions. Figure 3.38b) shows the same data set after background subtraction.

To represent the ON data, one calculates the squared angular difference $\theta^2$ between the reconstructed primary particle arrival direction and the source position. There are two possibilities to evaluate the background: First, one can use the background model of figure 3.37d to compute the background $\theta^2$ distribution at the source position. Second, one can

## 3.5 Analysis of Data Taken in the Wobble Mode

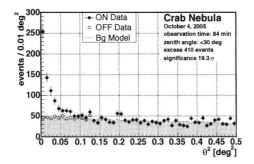

Figure 3.39: *Distributions of $\theta^2$ with respect to the source location (ON Data), the three background control positions (OFF data), scaled by a factor 1/3, and the calculated background model for Crab Nebula wobble observations (data set as in figure 3.36). A lower Size cut of 400 photoelectrons has been applied (analysis energy threshold about 280 GeV). For large values of $\theta^2$ the distribution of the ON data and the background model agree well, no systematic differences can be seen. For $\theta^2 < 0.04\,\text{deg}^2$ there is an excess of 410 $\gamma$-ray candidate events of the ON data over the OFF data, corresponding to a significance of 19.3 $\sigma$.*

use the $\theta^2$ distribution with respect to the three background control positions. Of course the latter distribution has to be truncated for larger $\theta^2$ values for which also signal events from the source would be included. Figure 3.39 shows as an example the $\theta^2$ distributions with respect to the source position (ON data), the three background control positions (OFF) data and the computed background model. The OFF data and the generated background agree well. Also the distribution of the ON data agrees well with the generated distribution from the background model for larger $\theta^2$ values. For $\theta^2 < 0.04\,\text{deg}^2$ there is an excess of 410 events, corresponding to a significance of 19.3 $\sigma$. In case of using a fit to the OFF data distribution one obtains a significance of 24.1 $\sigma$.

The use of three background control regions to evaluate the background to the ON data, yields a statistically better determination of the background in the Disp/$\theta^2$ analysis compared to the case of the Alpha analysis of the wobble data (see above). Therefore, for wobble data the Disp/$\theta^2$ analysis yields a better sensitivity than the Alpha analysis. Moreover, the anti-Alpha cut in the case of the Alpha analysis of the wobble data further reduces the effective area (the number of signal events) and thus the sensitivity of the Alpha analysis of the wobble mode data.

### 3.5.3 Comparison of the ON/OFF and Wobble Observation Modi

For the decision about the observation modus it is of vital interest to compare the achievable sensitivities with the ON/OFF and wobble mode observations. In the presented sample analysis (no cut optimization for the highest sensitivity) the sensitivity, measured in significance per square root ON observation time, is about the same for the ON/OFF as for the wobble observations. The small trigger region of the MAGIC camera causes an acceptance loss of up to 20% for wobble observations compared to ON observations. Increasing the trigger region of the camera in future upgrade projects would also increase the sensitivity in the wobble mode but have only negligible effects on the sensitivity for ON observations.

For the reliable background determination to the ON data, dedicated OFF data have to be taken in addition to the ON data, while the wobble observation mode allows the reliable background estimation from the wobble data alone (in case of a homogeneous camera acceptance). Taking further into account that some sources have a very particular sky brightness profile and changing atmospheric and telescope conditions, the wobble observation mode is becoming the favorite observation mode for steady state, point-like and slightly extended (see e.g. section 4.3) sources.

The observations of the sources HESS J1813-178 and HESS J1834-087, presented in sections 4.2 and 4.3, have been conducted in the wobble mode. The Galactic Center (see section 4.1) was observed in ON/OFF as well as in the wobble mode.

## 3.6 Basic Performance Parameters of the MAGIC Telescope

The most important performance parameters of a Cherenkov telescope are the flux sensitivity (section 3.6.1), the angular resolution (section 3.6.2), the energy threshold and the energy resolution (section 3.6.3). Further important parameters are the systematic uncertainties in the $\gamma$-ray flux determination (section 3.7). They are tightly linked to a precise knowledge of the optical parameters of the telescope and the atmosphere as well as the quality of the agreement between $\gamma$-ray candidate images and MC simulated $\gamma$-ray images, see section 3.6.4. For the analysis presented in this thesis the performance of the MAGIC telescope for observations under large ZAs and the reconstruction of extended sources are relevant and are presented in sections 3.6.5 and 3.6.6.

### 3.6.1 Sensitivity

The sensitivity of the MAGIC telescope is defined as the minimum detectable flux of a source in a given observation time and at a given significance level. Contrary to the (detection) significance calculated for the source observations, equation 3.13, for the sensitivity calculation a different (flux) significance definition is used:

$$S_\gamma = \frac{N_{\text{ON}} - N_{\text{OFF}}}{\sqrt{N_{\text{OFF}}}} \ . \tag{3.23}$$

Note, that in case of a small signal to background ratio $(N_{\text{ON}} - N_{\text{OFF}})/(N_{\text{ON}} + N_{\text{OFF}})$ both significance definitions (equations 3.13 and 3.23) give approximately the same result.

In general the sensitivity is defined as a $5\sigma$ significance detection for an observation time of 50 hours. Figure 3.40 shows the integral point source flux sensitivity of MAGIC as obtained from the MC simulation and the analysis of the Crab Nebula and other sources, after Teshima et al. (2005).

## 3.6 Basic Performance Parameters of the MAGIC Telescope

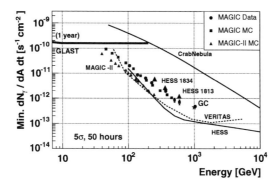

Figure 3.40: *MAGIC integral point source flux sensitivity as a function of the analysis energy threshold for low ZA observations ($0° \leq$ ZA $\leq 30°$) and a $\gamma$-ray spectrum of $dN/dE \propto E^{-2.6}$ for data and MC simulations at low ZA, after Teshima et al. (2005). The integral flux sensitivities of the analyses presented in sections 4.1 to 4.3 (large ZA, Galactic sources) are also indicated. In 50 hours a significant signal of a source with a few percent of the Crab Nebula flux can be observed.*

### 3.6.2 Angular Resolution

The angular resolution is defined as the sigma of a two-dimensional Gaussian, $\sigma_{2D}$, fitted to the brightness distribution $b(x, y)$ (photons per solid angle) of an optical or $\gamma$-ray point source in the center of the MAGIC camera: $b(x,y) \propto \exp\left(-\frac{x^2+y^2}{2\sigma_{2D}^2}\right)$ (Cortina 2005). Figure 3.41a shows the reconstructed $\gamma$-ray PSF using the Disp method as a function of the lower cut on the image parameter Size, obtained for the Crab Nebula. $\sigma_{2D}$ is around $0.1°$ with a weak dependence on the lower Size cut.

In case of VHE $\gamma$-rays the PSF contains the contributions of the instrument as well as the reconstruction accuracy using the Disp method (see section 3.3.2). Using the MAGIC automatic mirror control, the mirror positions are frequently readjusted to yield an optical PSF of $\sigma_{2D} = 0.040° - 0.045°$ (Garczarczyk 2006). The optical PSF is defined as the sigma of a two-dimensional Gaussian fitted to the optical brightness distribution of a star image in the MAGIC camera plane. Figure 3.41b shows an example for the brightness distribution of the optical image of a bright star (Vega) in the MAGIC camera plane.

### 3.6.3 Energy Threshold / Energy Resolution

The energy threshold for an observation is defined as the energy where the distribution of $\gamma$-ray excess events per energy interval reaches its maximum. The energy threshold depends on the telescope location, especially the height above the sea level, the telescope parameters like light collection area and quantum efficiency of the photomultipliers as well

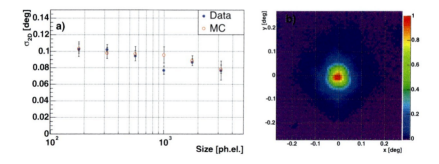

Figure 3.41: *Angular resolution: a) Sigma of a two-dimensional Gaussian fitted to the background subtracted sky map of γ-ray excess events from the Crab Nebula and MC simulated γ-ray showers as a function of a lower cut in Size. b) Brightness distribution (arbitrary units) of the optical image of a star in the MAGIC camera plane, a fit with a two-dimensional Gaussian yields a sigma of about 0.04°, figure adapted from (Garczarczyk et al. 2006). The size of an inner camera pixel is 0.1°.*

as the lower image Size cut and the applied γ/hadron separation cuts. Moreover, the energy threshold depends strongly on the observation zenith angle, see section 3.6.5.

The observations presented in this thesis were conducted at large zenith angles between 37° and 62° in an environment of high sky brightness from the Galactic plane. Therefore, conservative analysis cuts have been chosen not optimizing for the lowest possible analysis energy threshold. For a phenomenological parameterization of the analysis energy threshold as a function of the observation zenith angle, see section 3.6.5.

The relative energy resolution is on average about 25% and itself a function of the energy of the γ-ray, see section 3.2.6. The relatively coarse energy resolution requires an unfolding procedure for the determination of the source energy spectrum, see section 3.4.3.

### 3.6.4 MC - Data Comparison

The IACT method does not offer the possibility to evaluate the γ/hadron separation cut efficiency and the energy estimation resolution by means of test beams of VHE γ-rays of known energy. They can only be evaluated using MC simulations.

The γ-ray efficiency translates directly into the effective collection area used in the flux calculation (see section 3.4). Therefore, a good agreement between the camera images of γ-ray showers and the MC simulated ones is a necessary requirement for the reliable flux measurement. Every difference leads to a wrong estimate of the γ-ray efficiency and the effective collection area. In the standard analysis chain no MC simulations of the backgrounds (mostly hadron showers) are used as the simulation of hadronic reactions is more difficult compared to electromagnetic ones, see e.g. Cojocary et al. (2004).

In order to compare the MC simulated γ-ray shower images to the real ones, the dis-

## 3.6 Basic Performance Parameters of the MAGIC Telescope

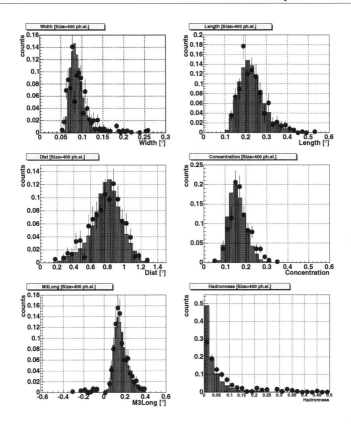

Figure 3.42: *Comparison of the image parameters Width, Length, Dist, Concentration, third moment along the long axis and Hadronness for MC $\gamma$-rays and $\gamma$-ray candidates measured from the Crab Nebula. A lower Size cut of 400 photoelectrons was applied. The real data are found to be in a reasonable agreement with Monte Carlo expectations.*

tributions of image parameters are studied. In principle, also the intercorrelation between the parameters would have to be compared. From the measured data the image parameter distributions of the $\gamma$-ray excess events can be obtained by subtracting the corresponding measured image parameter distributions for OFF from ON-source data after a suitable normalization (Majumdar et al. 2005). Apart from data quality cuts, only an extremely loose hadron suppression cut (rejecting just 2% of $\gamma$-ray showers while rejecting half of the background) has been applied to keep the number of events in the subtracted histograms high and hence to reduce the fluctuations in the resulting distributions. Moreover, this hadronness cut has to be loose to avoid biasing the image parameter distributions of the selected events. In the case of a strong hadronness cut, the selected events would by construction agree with the MC simulations. The same data quality and very loose hadronness cuts are also applied to the MC $\gamma$-ray sample. In filling the histograms of the image parameter distributions, the Monte Carlo generated $\gamma$-ray showers have been weighted to account for the deviation of the Crab spectrum from a pure power law towards low energies (Lucarelli et al. 2003).

Figure 3.42 displays the resulting distributions of the image parameters Width, Length, Size, Conc, M3Long and Hadronness for the observed $\gamma$-ray excess events from the Crab Nebula (minimum Size cut of 200 photoelectrons) and the MC simulated $\gamma$-rays. The real data are found to be in a reasonable agreement with Monte Carlo expectations. Nevertheless, there is a disagreement between data and MC beyond statistical errors in the hadronness distribution for hadronness values below 0.2. Using MC simulations to determine the $\gamma$/hadron separation cut efficiency this disagreement between data and MC simulations may introduce a systematic error in the reconstructed $\gamma$-ray flux (see section 3.7). This systematic effect can be minimized by using conservative hadronness cuts having a high $\gamma$-ray efficiency.

### 3.6.5 Observation at Large ZAs

For observations at large ZA the $\gamma$-ray initiated shower cascade develops higher up in the atmosphere and farther away from the Cherenkov telescope. The Cherenkov light illuminates a larger area on the ground, see figure 3.43. A telescope anywhere in this illuminated area (note: the effective detection area in the flux calculation is the area of the Cherenkov light cone perpendicular to the pointing direction of the telescope) will see the $\gamma$-ray shower, but with an accordingly reduced brightness. Thus the energy threshold for the observation of $\gamma$-ray showers increases with increasing ZA, but also the effective area (the detection rate) increases, see e.g. Konopelko et al. (1999) and Firpo (2006). Observations at large ZA (up to 60°) therefore allow one to measure the high-energy part of the spectrum with improved precision, see e.g. Krennrich et al. (1999); Tanimori (1994).

Figure 3.44a presents the analysis energy threshold as a function of the ZA. A conservative image cleaning (thresholds: 10 photoelectrons for core pixels and 5 photoelectrons for boundary pixels) and a conservative lower Size cut of 200 photoelectrons have been applied. The full line shows an empirical scaling law for the energy threshold:

$$E_{\text{threshold}}(\text{ZA}) = E_{\text{threshold}}(\text{ZA} = 0°)(\cos \text{ZA})^{-2.65} \ . \tag{3.24}$$

Figure 3.44b shows the corresponding effective collection area for $\gamma$-ray showers for energies above two times the energy threshold as a function of ZA. The full line shows an empirical scaling law for the effective collection area for the particular analysis chosen:

$$A_{\text{eff.}}(\text{ZA}) = A_{\text{eff.}}(\text{ZA} = 0°) + (\sin \text{ZA})^8 \cdot 7.4 \cdot 10^5 \, \text{m}^2 \ . \tag{3.25}$$

The Crab Nebula is a bright and steady source of VHE $\gamma$-rays, see e.g. Aharonian et al. (2004a); Hillas et al. (1998). From La Palma it can be observed under ZAs above about 7°. Therefore, the analysis at large zenith angles was developed and verified using Crab

## 3.6 Basic Performance Parameters of the MAGIC Telescope

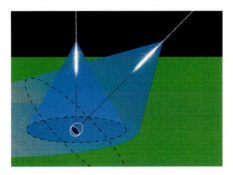

Figure 3.43: *For observations at large ZA the γ-ray initiated shower cascade develops higher up in the atmosphere and farther away from the Cherenkov telescope. The Cherenkov light illuminates a larger area on the ground. A telescope anywhere in this illuminated area will see the γ-ray shower, but with an accordingly reduced brightness. Thus the energy threshold for the observation of γ-ray showers increases with increasing ZA, but also the effective area (the detection rate) increases. Picture taken from Firpo (2006).*

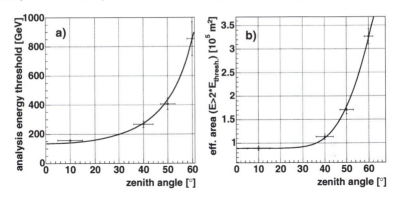

Figure 3.44: *a) Analysis energy threshold as a function of the ZA. A conservative image cleaning (thresholds: 10 photoelectrons for core pixels and 5 photoelectrons for boundary pixels) and a conservative lower Size cuts of 200 photoelectrons have been applied. The full line shows an empirical scaling law for the energy threshold, see text. b) The corresponding effective collection area for energies above two times the energy threshold as a function of ZA. The full line shows an empirical scaling law for the effective collection area, see text.*

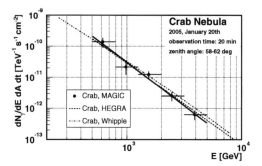

Figure 3.45: *Energy spectrum of the Crab Nebula observed at large zenith angles between 58° and 62°. Within errors it agrees with the previous measurements by the HEGRA collaboration (Aharonian et al. 2004a) and the spectral model fit to the Whipple data by Hillas et al. (1998). The full line shows a fit to the MAGIC data taking the full instrumental energy resolution into account, see text.*

Nebula data with a ZA around 60°. Figure 3.45 displays the reconstructed VHE $\gamma$-ray spectrum (dN$_\gamma$/(dE$_\gamma$dAdt) vs. true E$_\gamma$) of the Crab Nebula after correcting (unfolding) for the instrumental energy resolution. The full line shows the result of a forward unfolding procedure, see section 3.4.3. The result of the fit in the region 500 GeV $< E_{\text{estimated}} <$ 5 TeV is given by ($\chi^2$/n.d.f = 2.6/3):

$$\frac{\mathrm{d}N_\gamma}{\mathrm{d}A\mathrm{d}t\mathrm{d}E} = (3.0 \pm 0.6) \times 10^{-11} \left(\frac{E}{\text{TeV}}\right)^{-2.8\pm 0.3} \text{cm}^{-2}\text{s}^{-1}\text{TeV}^{-1}. \quad (3.26)$$

The given errors (1$\sigma$) are purely statistical. Within errors, the reconstructed energy spectrum of the Crab Nebula observed under large zenith angles agrees with the one reconstructed from observations under small zenith angles (see equation 3.23 and figure 3.31). The analysis energy threshold has increased from about 170 GeV at low ZA to about 850 GeV for a ZA of 60°. The measured Crab Nebula flux at large ZAs also agrees with previous measurements at small and large ZAs by the CANGAROO (Tanimori 1994), Whipple (Hillas et al. 1998) and HEGRA (Aharonian et al. 2004a) collaborations.

The point-source flux sensitivity obtained for large ZA observations of the Galactic Center (58° $\leq$ ZA $\leq$ 62°, see section 4.1) and the source HESS J1813-178 (47° $\leq$ ZA $\leq$ 54°, see section 4.2) is shown in figure 3.40.

### 3.6.6 Extended Sources

A $\gamma$-ray point source is reconstructed in the background subtracted $\gamma$-ray sky map with a brightness distribution $b(x,y) \propto \exp\left(-\frac{x^2+y^2}{2\sigma_{\text{PSF}}^2}\right)$, where $\sigma_{\text{PSF}}$ depends weakly on the energy, see section 3.6.2. In case that the $\gamma$-ray source itself has a two-dimensional Gaussian

## 3.7 Systematic Errors

profile with $\sigma_{\text{source}}$ than the reconstructed brightness distribution follows a two-dimensional Gaussian with standard deviation $\sigma_{\text{total}}$:

$$\sigma_{\text{total}} = \sqrt{\sigma_{\text{PSF}}^2 + \sigma_{\text{source}}^2} \ . \tag{3.27}$$

For arbitrary source $\gamma$-ray brightness morphologies the reconstructed skymap brightness distribution is the convolution of the source morphology with the instrument PSF.

Accordingly, also the width of the $\theta^2$ distribution increases with increasing intrinsic width of the source. Figure 3.46a displays $\theta^2$ distributions for MC simulated $\gamma$-ray showers for sources with different intrinsic extensions, while figure 3.46b shows the corresponding distributions for the image orientation angle Alpha. The distributions become broader with increasing source size. The necessary larger signal integration regions contain more background and thus the sensitivity for the reconstruction of extended sources is reduced. To determine the cut efficiency of the $\theta^2$ and Alpha cut, one has to use dedicated MC simulations which simulate the source intrinsic brightness profile. For a source extension of 0.4° and larger a considerable part of the $\gamma$-ray signal leaks into the ON/OFF normalization region at large $\theta^2$ and Alpha values. Thus the normalization of the ON and OFF data becomes problematic. For further discussion about the reconstruction of extended sources see e.g. Aharonian et al. (1994).

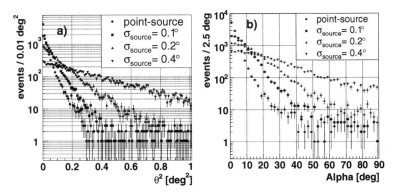

Figure 3.46: *a) Distributions for the $\theta^2$ parameter for MC simulated $\gamma$-ray showers for sources with different intrinsic extension. b) Corresponding distributions for the image orientation angle Alpha. A minimum Size cut of 200 photoelectrons has been applied.*

## 3.7 Systematic Errors

In addition to statistical errors the $\gamma$-ray flux and source position/morphology are affected by systematic uncertainties. There are three dominating types of systematic errors of the $\gamma$-ray flux measurement:

1. Errors in the reconstruction of the $\gamma$-ray energy
2. Errors in the calculation of the effective collection area
3. Errors in the determination of the background to the signal events.

The main sources for errors in the reconstruction of the $\gamma$-ray energy are:

- **Cherenkov light production**: The production and development of an electromagnetic shower are very well understood (see e.g. Cojocary et al. (2004) and references therein). Also the production of Cherenkov light in an electromagnetic air shower is well understood and can be modeled with high precision (Heck et al. 1998). The corresponding systematic error on the reconstructed $\gamma$-ray energy should not exceed 1%.

- **Atmospheric model** used in the MC simulations: In the MC simulations a standard model of the atmosphere (so-called standard US atmosphere, see section 2.2.8) has been used. It does not take into account the microclimate conditions at La Palma (subtropic island climate) as well as seasonal variations. This can affect the reconstructed $\gamma$-ray energy by about 15% (Bernlöhr 2000).

- **Weather conditions**: The general weather conditions at La Palma are monitored, but the atmospheric extinction will only be measured in future with a Lidar system, see e.g. Merck (2004). Thin clouds, moisture and dust may cause a reduced atmospheric transmission. This leads to an uncertainty in the reconstructed image Sizes and reconstructed energies by up to 10%.

- **Light losses** in the optical system consisting of mirrors, the camera entrance window and the Winston cone light concentrators: The overall light losses are adjusted in the MC simulations to dedicated measurements and monitored by the muon calibration (see section 3.2.1). However, there may be errors in the simulated light distribution on the camera (e.g. the size of the PSF and halos of the focussed spot on the camera). The systematic error on the reconstructed energy may be 10%.

- **PMT quantum efficiency**: the absolute light-to-photoelectron conversion efficiency of the PMTs (wavelength dependent) can be measured using the PIN diode or blind pixel calibration (see section 3.2.1) for three different light wavelengths. The systematic error of the PIN diode calibration method has been estimated to be about 8% by Gaug et al. (2005).

Adding these five errors on the $\gamma$-ray energy in quadrature, one obtains a total systematic error on the reconstructed $\gamma$-ray energy of 22%. Generally, the differential $\gamma$-ray spectra are steeply falling with $\gamma$-ray energy and can be fitted by a power law $dN_\gamma/dE \propto E^{-\alpha}$. The error of the integral flux $\Phi(E > E_0) = dN_\gamma(E > E_0)/dAdt$ above an energy $E_0$ is then given by:

$$\frac{\Delta \Phi}{\Phi} = (\alpha - 1)\frac{\Delta E}{E} \ . \tag{3.28}$$

## 3.7 Systematic Errors

For a differential $\gamma$-ray spectral index of $\alpha = 2.0$ the 22% systematic error in the reconstructed $\gamma$-ray energy translates in a 22% systematic error of the integral flux. $\alpha = 2.6$ (as for the Crab Nebula) leads to 35% systematic error of the integral flux.

The second type of systematic errors is due to uncertainties in the determination of the effective collection area. The main error sources are:

- **Camera acceptance**: In the MC simulations a camera with homogeneous acceptance is simulated. Defect PMTs and trigger inefficiencies may introduce a systematic error of about 5% of the flux level.

- **$\gamma$/hadron separation efficiency**: It is determined by applying the $\gamma$/hadron separation algorithm to MC simulated $\gamma$-rays. Differences between the real and simulated images of $\gamma$-ray showers (partly due to differences of the simulated and real optical point spread function of the telescope and the earth's magnetic field), see section 3.6.4, may introduce a systematic error up to 10%, which may depend on the $\gamma$-ray energy.

- **Tracking errors**: The starguider allows to correct for relative mis-pointings but has not yet been calibrated to the absolute position in the sky (see section 3.3.2). A wrong tracking position leads to broader Alpha and $\theta^2$ distributions and a reduced efficiency of the chosen signal cut in Alpha or $\theta^2$ (see section 3.3). This effect may result in a few percent error of the flux level.

The third type of systematic errors is related to the determination of the $\gamma$-ray signal by means of the Alpha or Disp/$\theta^2$ analysis (see section 3.3). The main error sources are:

- **Star field**: the star field of the ON data is in general different from the one of the OFF data. Also for wobble data there is a translation of the star field in the camera when changing the tracking positions (see sections 2.3 and 3.5). Bright stars and gradients in the sky brightness may affect the shape of the Alpha and $\theta^2$ distributions used to evaluate the $\gamma$-ray signal. This effect is most important for small image Sizes. For the analysis presented here high image cleaning cuts (see section 3.2.3) have been used minimizing the star field effect. The systematic error on the flux is estimated to be 5% and may depend on the $\gamma$-ray energy.

- **Camera inhomogeneity**: To estimate the background in the wobble observation mode, several background control regions are used, which are located rotationally symmetric in the camera (see section 3.5). In case of an inhomogeneous camera (different acceptance for signal and background) systematic errors in the background determination arise. The error is proportional to the signal to background ratio and may therefore depend on the $\gamma$-ray energy.

- **$\gamma$-ray sources in OFF date**: Especially for the observation of galactic OFF data there is a certain possibility that the OFF data contain a $\gamma$-ray source. For example, recent observations by Aharonian et al. (2006c) show that there is diffuse $\gamma$-ray emission from the Galactic plane $\pm 3°$ around the galactic center. For observations of the Galactic Center the systematic error is estimated to be 5% in the flux level.

Adding the systematic errors in quadrature, one obtains a systematic error in the integral flux level of 28% (for $\alpha = 2.0$) to 37% (for $\alpha = 2.6$). Nevertheless, some of the systematic errors cause an underestimation of the flux level such that the real flux level may be up to 60% larger (linear addition of the systematic errors) than the reconstructed one. Only few of the systematic errors depend on the $\gamma$-ray energy (like the effect of the star field) and introduce an error in the slope of the differential $\gamma$-ray spectrum. To exactly evaluate the systematic error in the spectral slope extensive MC simulation studies are necessary. Based on the experience from previous and other similar Cherenkov telescopes (see e.g. Aharonian et al. (2006a)) the systematic error in the spectral slope has been assumed to be 0.2.

The systematic errors of the position/morphology determination are due to the following effects:

- **telescope positioning** error: The used 14 bit shaft encoders allow a telescope positioning with an accuracy of 2', see section 2.2.2.

- **starguider**: The starguider was not calibrated to absolute sky positions for the data analyzed in this thesis, see section 3.3.2. Therefore, it cannot correct for the systematic telescope pointing error.

- **Star field**: The background to the $\gamma$-ray sky maps is generated from the measured telescope acceptance, see section 3.3.2. Bright stars and gradients in the sky brightness around the $\gamma$-ray source may affect the background distribution in the sky. This effect depends on the signal to background ratio of the source and on the chosen lower cut on the image Size.

The total systematic error in the source position determination is dominated by the 2' telescope pointing accuracy.

# Chapter 4

# Observation of VHE γ-Rays from Galactic Sources

In this chapter the observations of VHE γ-rays with the MAGIC telescope from three galactic sources are discussed: the Galactic Center (section 4.1), HESS J1813-178 (section 4.2) and HESS J1834-087 (section 4.3). The source positions, extensions and the energy spectrum of the VHE γ-rays are determined. Possible VHE γ-ray flux variations with time are studied. The results are put in the perspective of multiwavelength observations and models for the multiwavelength emission.

## 4.1 Observation of VHE γ-Rays from the Galactic Center

This section starts with an introduction to the observation of the Galactic Center in γ-rays (section 4.1.1), followed by a description of the observational technique used (section 4.1.2) and the procedure implemented for the data analysis (section 4.1.3). In section 4.1.4 the results are discussed in the perspective of different models proposed for the production of the VHE γ-rays. Finally, section 4.1.5 contains the conclusions. The results of this analysis as presented in this section have been published in Albert et al. (2006b).

### 4.1.1 Introduction

The Galactic Center (GC) region contains many remarkable objects which may be responsible for high-energy processes generating γ-rays (Aharonian & Neronov 2005; Atoyan & Dermer 2004), see also section 1.5.6. The GC is rich in massive stellar clusters with up to 100 OB stars (Morris & Serabyn 1996), immersed in a dense gas. Also, young supernova remnants can be found, e.g. Sgr A East, and nonthermal radio arcs (LaRosa et al. 2000). The dynamical center of the Milky Way is associated with the compact radio source Sgr A*, which is believed to contain a massive black hole of about $(3-4) \cdot 10^6 M_\odot$ (Morris & Serabyn 1996; Schödel et al. 2002). Within a radius of 10 pc around the GC there is a mass of about $3 \cdot 10^7 M_\odot$ (Schödel et al. 2002; Genzel et al. 2003).

EGRET has detected a strong source (3 EG J1745-2852) near the GC. Hartman et al. (1999) and Mayer-Hasselwander et al. (1998) find the source position to be consistent with the GC position within the EGRET error circle of 0.2° radius. However, an independent analysis of the EGRET data (Hooper & Dingus 2005) indicates a point source whose position is different from the GC at a confidence level beyond 99.9 %. This was recently confirmed by Pohl (2005). 3 EG J1745-2852 has a broken power law energy spectrum extending up to at least 10 GeV. Below the break of the energy spectrum at a few GeV the spectral index is 1.3, see figures 1.9 and 1.11. Assuming a distance of the GC of 8 kpc, the $\gamma$-ray luminosity of this source is very large, $2.2 \cdot 10^{37}$ erg/s, which is equivalent to about 10 times the $\gamma$-ray flux from the Crab Nebula.

In VHE $\gamma$-rays the GC has been observed by the CANGAROO (Tsuchiya et al. 2002, 2004), VERITAS (Kosack et al. 2004) and HESS collaborations (Aharonian et al. 2004b). The energy spectra as measured by these groups show substantial differences both in the flux level and in the spectral slope. This might be due to different sky integration regions of the signal or to a source variability at a time-scale of about one year, or to inter-calibration problems.

### 4.1.2 Observations with the MAGIC Telescope

At La Palma, the GC ((RA, Dec) = $(17^h 45^m 36^s, -28°56')$) culminates at about 58° zenith angle (ZA). The star field around the GC is non-uniform. In the region RA > $RA_{GC} + 4.7^m$ the star field is brighter. Within a distance of 1° from the GC there are no stars brighter than 8 mag. Figure 4.1 shows the star field around the GC.

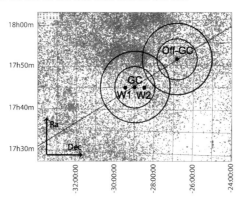

Figure 4.1: *The star field around the Galactic Center. The 2 sets of big circles correspond to distances of 1° and 1.75° from the GC and OFF-GC, respectively. The wobble positions W1 and W2 are given by the full circles. The full black line represents the Galactic Plane.*

The MAGIC observations were carried out in the ON/OFF mode as well as in the false-source tracking (wobble) mode (see section 2.3). The sky directions (W1, W2) to be tracked

## 4.1 Observation of VHE γ-Rays from the Galactic Center

in the wobble mode are chosen such that in the camera the star field relative to the source position (GC) is similar to the star field relative to the mirror source position (anti-source position): W1/W2 = (RA$_{GC}$, Dec$_{GC}$ ± 0.4°). During wobble mode data taking, 50% of the data is taken at W1 and 50% at W2, switching between the two positions every 20 minutes. Dedicated OFF data have been taken, with a sky field similar to that of the ON region. The OFF region is centered at the Galactic Plane, GC$_{OFF}$ = ($17^h51^m12^s$, $-26°52'00''$), see figure 4.1. OFF data was taken during the same night, directly before and after the ON observations under the same weather conditions and with the same hardware setup. The ON and OFF observations were conducted such that they have similar ZA ranges and distributions, although the OFF source position has a slightly different declination than the GC. After initial observations in September 2004 the GC was observed for a total of about 24 hours in the period May-July 2005. Table 4.1 summarizes the data taken.

| Period | date | ZA [°] | time [h] | events [$10^6$] | observation mode |
|--------|------|--------|----------|-----------------|------------------|
| I | Sep. 2004 | 62-68 | 2 | 0.8 | ON |
| II | May 2005 | 58-62 | 7 | 2.8 | wobble |
| III | Jun./Jul. 2005 | 58-62 | 17/12 | 6.4/5.0 | ON/OFF |

Table 4.1: *Data set per observation period of the GC. The column "time" states the effective observation time, the column "events" states the number of events after image cleaning.*

### 4.1.3 Data Analysis

The data analysis has been carried out using the standard MAGIC analysis and reconstruction software (Bretz & Wagner 2003), the first step of which involves the signal reconstruction using the digital filter and the calibration using the F-factor method (see sections 3.1 and 3.2.1). After calibration, image cleaning tail cuts of 10 photoelectrons for core pixels and 5 photoelectrons for boundary pixels have been applied (see section 3.2.3). These tail cuts are accordingly scaled for the larger size of the outer pixels of the MAGIC camera. The camera images are parameterized by image parameters, see section 3.2.4.

In the analysis presented here, a custom implementation (Bock et al. 2004; Hengstebeck 2003) of the Random Forest (RF) method (Breiman 2001) was applied for the γ/hadron separation and the energy estimation, as described in section 3.2.5. In the RF method several independent decision trees (series of cuts) are applied to each event. By combining the results of the individual decision trees, two parameters are calculated: the estimated energy and the parameter hadronness, which is a measure of the probability that the event is not γ-ray like.

The trees of the RF are generated by means of training samples for the different classes: A sample of Monte Carlo (MC) generated γ-ray showers were used to represent the γ-ray showers together with about 1% randomly selected events drawn from the measured OFF-data as background showers. The MC γ-ray showers were generated between 58° and 68° ZA with energies between 10 GeV and 30 TeV. For the analysis of the September 2004 data set the RF cuts were determined using a sub-set of Galactic OFF data as background. The source-position independent image parameters (see section 3.2.4) Size, Width, Length

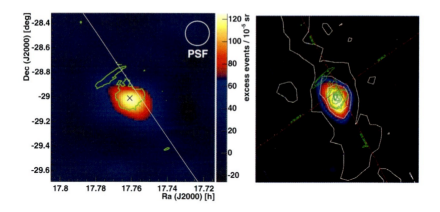

Figure 4.2: *Sky maps of γ-ray candidates (background subtracted) in the direction of the Galactic Center for Size ≥ 300 photoelectrons (corresponding to an energy threshold of about 1 TeV). A Gaussian folding has been applied. The left sky map is overlayed with green contours (0.3 Jy beam$^{-1}$) of 90 cm VLA (BCD configuration) radio data (LaRosa et al. 2000). The black cross shows the position of Sgr A* and the white line shows the Galactic Plane. The right sky map is overlayed with green contours of the same radio data and white contours showing the VHE γ-ray HESS data (Aharonian et al. 2004b). The MAGIC and HESS data match well concerning the source location as well as the source morphology.*

and Concentration, as well as the source-position dependent parameter Dist and the third moment of the photoelectrons distribution along the major image axis, were selected to parameterize the shower images. The γ-ray sample is defined by selecting showers with a hadronness below a specified value. An independent sample of MC γ-ray showers was used to determine the efficiency of the cuts.

The analysis at high zenith angles was developed and verified using Crab data with a ZA around 60°. The reconstructed Crab energy spectrum was found to be consistent within $1\sigma$ statistical uncertainty with measurements at small zenith angles and other existing measurements, see section 3.6.5.

For each event the arrival direction of the primary particle in sky coordinates is estimated by the Disp-method (see section 3.2.7): The arrival direction is assumed to be on the major axis of the Hillas ellipse that fits the shower image in the camera, at a certain distance (Disp) from the image center of gravity. Figure 4.2 shows the background subtracted sky map of γ-ray candidates from the GC region (observation periods II/III). It is folded with a two-dimensional Gaussian with a standard deviation of 0.072° and height one (the MAGIC γ-ray PSF is ∼ 0.1°, see section 3.6.2). A lower Size cut of 300 photoelectrons has been applied, corresponding to an energy threshold of about 1 TeV. The

## 4.1 Observation of VHE γ-Rays from the Galactic Center

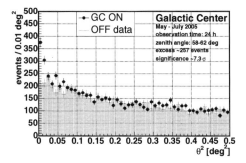

Figure 4.3: *Distributions of $\theta^2$ values for the Galactic Center and OFF data, see text, for Size $\geq$ 300 photoelectrons (corresponding to an energy threshold of about 1 TeV).*

sky map is overlayed with contours (0.3 Jy beam$^{-1}$) of 90 cm radio data, obtained from the Very Large Array[1] (VLA) of radio telescopes in BCD configuration from LaRosa et al. (2000). The brightest non-central source is the Arc. The excess is centered at (RA, Dec) = ($17^h45^m20^s$, -29°2′) (J2000 coordinates). The present systematic pointing uncertainty is estimated to be 2′, see section 2.2.2.

Figure 4.3 shows the distribution of the squared angular distance, $\theta^2$, between the reconstructed shower direction and the nominal GC position (corresponding to figure 4.2) and the corresponding OFF data (see sections 3.3.2 and 3.5.2) for the observation periods II/III. The observed excess in the direction of the GC has a significance of $7.3\sigma$ ($\theta^2 \leq 0.02 \deg^2$). For large values of $\theta^2$ the distributions for ON and OFF data agree well. The source position and the flux level are consistent with the measurement of HESS (Aharonian et al. 2004b) within errors.

The VHE γ-ray source G 0.9+0.1 (Aharonian et al. 2005b), a composite SNR, is located inside the MAGIC field-of-view. Figure 4.4 shows the distribution of the squared angular distance, $\theta^2$, between the reconstructed shower direction and the nominal position of G 0.9+0.1. The background has been determined according to section 3.3.2. For $\theta^2 \leq 0.02 \deg^2$ there is a small excess of about 3.2 standard deviations. Nevertheless, the background determination for a source as far as 1° offset from the camera center may be subject to systematic uncertainties (see section 3.7). The small excess is consistent with the low flux reported by Aharonian et al. (2005b) and the given observation time with the MAGIC telescope.

For the determination of the energy spectrum, the RF was trained including the source dependent image parameters Dist and third moment of the pixel charge distribution with respect to the nominal excess position. For the spectrum determination only the largest data set (period III) was used. The cut on the hadronness parameter (50% γ-ray efficiency corresponding to an effective area of about 250000 m$^2$) resulted in about 500 excess events

---
[1] For further information see http://www.vla.nrao.edu/

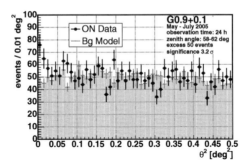

Figure 4.4: *Distributions of $\theta^2$ values for the composite SNR G 0.9+0.1, which is in the field of view of the Galactic Center, and a background model, see text, for Size $\geq$ 300 photoelectrons (corresponding to an energy threshold of about 1 TeV).*

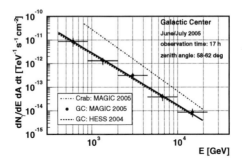

Figure 4.5: *Reconstructed VHE $\gamma$-ray energy spectrum of the GC (statistical errors only). The full line shows the result of a power-law fit to the data points taking the full instrumental energy resolution into account. The dashed line shows the result of the HESS collaboration (Aharonian et al. 2004b). The dot-dashed line shows the energy spectrum of the Crab Nebula as measured by MAGIC (Wagner et al. 2005). The horizontal bars indicate the bin size in energy, the marker is placed in the bin center on a logarithmic scale.*

## 4.1 Observation of VHE γ-Rays from the Galactic Center

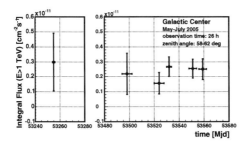

Figure 4.6: *Light curve: Reconstructed integral VHE γ-ray flux above 1 TeV as a function of time. The vertical error bars show 1σ statistical errors. The horizontal bars show the observation periods. The data are consistent with a steady emission within errors.*

with a minimum Size of 200 photoelectrons (corresponding to an energy threshold of about 700 GeV). Figure 4.5 shows the reconstructed VHE γ-ray energy spectrum of the GC after the unfolding of the instrumental energy resolution, see section 3.4.3. The horizontal bars indicate the bin size in energy, the marker is placed in the bin center on a logarithmic scale. The full line shows the result of a forward unfolding procedure as described in section 3.4.3: A simple power law spectrum is assumed for the true differential γ-ray flux. The parameters of the power law are determined by fitting the predicted differential flux to the measured energy spectrum. The result of the fit is given by ($\chi^2$/n.d.f = 5.1/5):

$$\frac{\mathrm{d}N_\gamma}{(\mathrm{d}A\mathrm{d}t\mathrm{d}E)} = (2.9 \pm 0.6) \times 10^{-12}(E/\mathrm{TeV})^{-2.2\pm 0.2}\ \mathrm{cm}^{-2}\mathrm{s}^{-1}\mathrm{TeV}^{-1}\ .$$

The quoted errors (1σ) are purely statistical. The systematic error is estimated to be 30% in the integral flux level and 0.2 in the spectral index, see section 3.7. The measured flux level as well as the spectral slope agree within the 1σ level with the HESS observations performed two years earlier. The measured spectral slope differs significantly from the value $-4.6 \pm 0.5$ given in the original publication of the CANGAROO collaboration (Tsuchiya et al. 2004). The CANGAROO collaboration has revised the error estimation of the Galactic Center data and quote a spectral slope of $-4.6^{+1.2}_{-5.0}$ (Katagiri et al. 2005).

Figure 4.6 shows the measured integral VHE γ-ray flux above 1 TeV as a function of time given as Modified Julian Date (MJD). The OFF data of all observation periods are used together to determine the background to the ON data for each time bin. This results in some correlation between the measured fluxes in the different time bins. Moreover, different observation modi may result in different systematic errors. The flux level is steady within errors in the time-scales explored by these observations.

## 4.1.4 Discussion

The observations of TeV γ-rays from the GC confirm that this is a very important region for high energy processes in the Galaxy. Many different objects, able to accelerate particles above TeV energies, are expected there (see section 1.5.6):

- the compact radio source and black hole candidate Sgr A*

- a possible AGN-like relativistic jet originating from the spinning GC black hole

- the young SNR Sgr A East

- the pulsar wind nebula candidate G359.95-0.04

- the diffuse central pc region

- the non-thermal radio filaments (Pohl 1997)

- the central part of the Dark Matter halo.

In the following, some of these objects are discussed as possible sources of the observed VHE γ-radiation.

### 4.1.4.1 Black Hole Candidate Sgr A*

The most likely source is considered to be the massive black hole identified with Sgr A* due to the directional coincidence. A blazar-like relativistic jet originating from the spinning GC black hole might produce TeV γ-rays (Falcke et al. 1993; Falcke & Markoff 2000). Due to an unfavorable orientation of the jet axis, the predicted flux of this model is lower than the observed VHE γ-ray flux.

Atoyan & Dermer (2004) proposed that electrons are accelerated to sufficiently high energies at the termination shock of the sub-relativistic wind from the central part of the advection dominated accretion flow onto the GC black hole, in analogy to the pulsar wind nebulae. The VHE γ-rays are produced by inverse Compton upscattering (see section 1.2.2) of the ambient submillimeter photons of the accretion flow (Atoyan & Dermer 2004). Liu et al. (2006) suggest that protons are energized by the process of stochastic acceleration. The observed VHE γ-rays are then produced by the decay of $\pi^0$s produced in collisions of the high energy protons with the dense ambient matter, see section 1.2.1. In both cases only the VHE γ-ray source can be described, but not the high flux of the EGRET source 3EG J1746-2851. This is consistent with the recent determination of the position of this EGRET source by Hooper & Dingus (2005) and Pohl (2005). They found that the position is different from the GC at the 99.9% level. Other scenarios for the γ-ray production in the vicinity of Sgr A*, have also been found to be consistent with the TeV observations but not with the GeV observations (Aharonian & Neronov 2005).

It is generally expected that γ-rays produced in such compact source models should show relatively fast variability like the variability observed in the infrared (Genzel et al. 2003) and X-rays (Baganoff et al. 2001, 2003; Porquet et al. 2003). The VHE γ-ray fluxes observed by the HESS telescope in 2003/2004 and by the MAGIC telescope agree within

## 4.1 Observation of VHE γ-Rays from the Galactic Center

errors. Also the MAGIC and HESS observations themselves, extending over a few months up to a year, rather suggest a stable source on a year time scale. However, the γ-ray flux above 2.8 TeV (3.7σ significance) reported by the VERITAS collaboration during the extended period from 1995 through 2003 is a factor ∼ 2 larger (Kosack et al. 2004) and the spectral slope reported by the CANGAROO collaboration is significantly steeper (Tsuchiya et al. 2004).

The present data are not sufficient to prove or reject Sgr A* as source of the VHE γ-rays. A more accurate determination of the position and possible extension of the VHE γ-ray source is necessary. A further test is the search for simultaneous flux variability in infrared, X-rays and VHE γ-rays.

### 4.1.4.2 Candidate PWN G359.95-0.04

The X-ray nebula G359.95-0.04 was discovered in deep Chandra X-ray observations of the Galactic Center (Wang et al. 2006). It lies at a projected distance of about 10" (corresponding to 0.3 pc at a distance of 8 kpc) to SgrA*. The nebula exhibits a cometary morphology with a projected size of $0.07 \times 0.3$ pc. The Chandra data reveal a softening of the X-ray spectral index with distance from the "head" of the nebula, which is a possible signature of cooling of electrons away from the accelerator. Wang et al. (2006) have suggested, that the head of the nebula contains a young pulsar and that G359.95-0.04 is likely a ram-pressure confined PWN. However, no point-like or extended source is observed at the position of G359.95-0.04 in the 6 cm radio band (private communication of Yusef-Zadeh, cited by Hinton & Aharonian (2006)).

G359.95-0.04 lies within the 68% confidence error circle of the VHE source at the Galactic Center of both the MAGIC and HESS measurements (Hinton & Aharonian 2006). Assuming a magnetic field of 120 $\mu$G, Hinton & Aharonian (2006) conclude that the same electron population, which produces the X-rays of G359.95-0.04 as synchrotron radiation in the ambient magnetic field may produce the VHE γ-rays by inverse Compton scattering (see section 1.2.2) of the high ambient far infrared background radiation at the GC. The radio upper limit requires a low energy cut-off of the electron distribution. In case that G359.95-0.04 is not responsible for the VHE γ-radiation this would put a lower limit of 120 $\mu$G to the average magnetic field in this region.

These IC models for the VHE γ-radiation predict a peak in the spectral energy distribution ($E_\gamma^2 dN_\gamma/dE_\gamma$) just below 100 GeV. Therefore, lower energy γ-ray data at a few GeV can decisively test this model. Also the VHE γ-ray emission from this source would be slightly (0.3 pc, corresponding to 8") offset from Sgr A* and expected to be stable in time.

### 4.1.4.3 SNR Sgr A East

The observed γ-ray emission may also come from the supernova remnant Sgr A East. This is a mixed morphology SNR (radio shell filled with an X-ray emitting nebula) with an age of about $10^4$ years (Maeda et al. 2002). Its elliptical structure is elongated along the Galactic Plane with a major axis of about 10.5 pc and a center displaced from the Galactic Center, Sgr A*, by about 2.5 pc (corresponding to 1' at a distance of 8 kpc) in projection towards higher right ascensions, see figure 1.8. The total energy release of Sgr A East is estimated to be between an average supernova value of about $10^{51}$ erg (Maeda et al. 2002) and about $4 \times 10^{52}$ erg (Mezger et al. 1989). The SNR is located in a strongly magnetized environment (about 180 $\mu$G) of high density (a hydrogen atom density of up to about $10^3$ cm$^{-3}$). In the interaction region with the molecular cloud M $-0.02$ $-0.07$ the matter density can be as high as $10^5$ cm$^{-3}$ and the magnetic field may be a few mG (Coil & Ho 2000). The high magnetic field together with a perpendicular shock geometry may provide a very efficient proton acceleration in the shell of the SNR (Crocker et al. 2005) up to extremely high energies of about $10^{18}$ eV.

Crocker et al. (2005) recognize that the spectra of the EGRET source (between 100 MeV and 10 GeV) and the VHE $\gamma$-ray source are incompatible (see figures 1.9 and 1.11) in case of hadronic emission. However, the EGRET GeV source may be offset by about 0.2° from the GC (Hooper & Dingus 2005; Pohl 2005). Therefore, Crocker et al. (2005), see also Fatuzzo & Melia (2003); Grasso & Maccione (2005), argue that the GeV and TeV $\gamma$-ray emission come from different sites of the shell of Sgr A East. The very high density environment may provide enough target material to produce the observed $\gamma$-ray luminosities within the given energy budget of the SNR. In this case also the VHE $\gamma$-ray source may be spatially offset from Sgr A* and it may be somewhat extended. The VHE $\gamma$-ray flux would be steady in this model.

Moreover, if in the SNR shell protons are accelerated to energies above $10^{18}$ eV they may produce neutrons (and neutrinos) of similar energy. The neutrons above $10^{18}$ eV may travel to the earth before they decay and would lead to a cosmic ray anisotropy in the direction of the GC (Crocker et al. 2005; Grasso & Maccione 2005). Corresponding claims for cosmic ray anisotropies were made by the AGASA and SUGAR experiments (Bossa et al. 2003), but recently the Auger collaboration reported not to see such an anisotropy (Letessier-Selvon et al. 2005).

To test the hadronic production of the VHE $\gamma$-rays in Sgr A East a more accurate determination of the position and possible extension of the VHE $\gamma$-ray source as well as possible flux variations with time are necessary. The observation of neutrinos or ultra high energy neutrons from Sgr A East would prove this model.

#### 4.1.4.4 The Diffuse Central pc Region

An extended $\gamma$-ray emission might originate from the interaction of relativistic particles with the soft radiation and matter of the central stellar cluster around the GC. The 95% confidence upper limit to the VHE $\gamma$-ray source size is 3' (corresponding to 7 pc at a distance of 8 kpc) (Aharonian et al. 2004b). This is compatible with the whole diffuse innermost few pc region being the source of the VHE $\gamma$-rays.

There are different mechanisms proposed for the acceleration of the very high energy particles:

- a very energetic pulsar (Bednarek 2002)
- a $\gamma$-ray burst source (Biermann et al. 2004)
- shocks in the winds of the massive stars (Quataert & Loeb 2005)

## 4.1 Observation of VHE γ-Rays from the Galactic Center

- the capture of a red giant star by the central supermassive black hole (Lu et al. 2006).

If the TeV γ-rays are produced by electrons, up-scattering the infrared photons from dust, heated by the UV stellar radiation (Quataert & Loeb 2005), then the γ-ray energy spectrum should significantly steepen between $\sim 0.1 - 1$ TeV. The Klein-Nishina effect suppresses the scattering of UV photons for electrons above TeV energies. Instead, the measured spectrum of VHE γ-rays can be well described by a single power law between 500 GeV and $\sim 20$ TeV, consistent with the results of the HESS collaboration (in the energy range $\sim 0.2 - 10$ TeV) (Aharonian et al. 2004b).

The VHE γ-rays may also be produced by hadronic interactions (see section 1.2.1). In order to reproduce the observed γ-ray spectrum, hadrons should have energies of at least $2 \times 10^3$ TeV. Such hadrons diffuse through the region of the TeV source ($< 7$ pc, Aharonian et al. (2004b)) on a time scale of the order of $10^4$ years, assuming an average magnetic field strength in this region of 100 $\mu$G and the Bohm diffusion coefficient. Thus, the natural source of relativistic hadrons seems to be the supernova remnant Sgr A East (see section 4.1.4.3 above) or the energetic pulsar created in the supernova explosion (Crocker et al. 2005; LaRosa et al. 2005; Bednarek 2002). However, this relatively young source of relativistic hadrons cannot be identified with the last γ-ray burst in the center of our Galaxy if it appeared $\sim 10^6$ years ago (Biermann et al. 2004).

In addition to the point-like source of VHE γ-rays at the Galactic Center, diffuse γ-ray emission along the Galactic Plane ($|b| < 0.2°$ and $|l| < 1.5°$) was reported by Aharonian et al. (2006c). The observed γ-ray flux is essentially proportional to the gas density. The observed energy spectrum of the γ-radiation has a similar hard power law slope of about $-2.3$ as the point source at the GC, harder than the slope of about $-2.7$ of the cosmic rays observed at Earth. Therefore, Aharonian et al. (2006c) conclude, that the Galactic Center clouds are illuminated by a cosmic-ray accelerator near the GC. The age of this object was estimated to be some $10^4$ years. The cosmic ray accelerator may well be the same object which is visible as point source of VHE γ-rays at the Galactic Center.

#### 4.1.4.5 Annihilation of Dark Matter Particles

The GC could also be the brightest source of VHE γ-rays from particle Dark Matter annihilation (Prada et al. 2004; Hooper et al. 2004; Flix 2005; Bartko et al. 2005), see section 1.7. Most SUSY Dark Matter scenarios lead to a cut-off in the γ-ray energy spectrum below 10 TeV. The observed γ-ray energy spectrum extends up to 20 TeV. Thus most probably the main part of the observed VHE γ-radiation is not due to Dark Matter annihilation (Horns 2004). However, an extended γ-ray source due to Dark Matter annihilation peaking in the region 10 GeV to 100 GeV (Elsässer & Mannheim 2005) cannot be ruled out yet.

### 4.1.5 Concluding Remarks

The MAGIC observations confirm the VHE γ-ray source at the Galactic Center. The measured flux is compatible with the measurement of the HESS collaboration (Aharonian et al. 2004b) within errors, but not with the measurements of the CANGAROO and VERITAS collaborations (Tsuchiya et al. 2004; Kosack et al. 2004). The VHE $\gamma$-ray emission does not show any significant time variability; the MAGIC measurements rather affirm a steady emission of $\gamma$-rays from the GC region. The excess is point-like, its location is consistent with SgrA*, G359.95-0.04 as well as SgrA East.

The nature of the source of the VHE $\gamma$-rays has not yet been identified. The main part of the observed VHE $\gamma$-radiation is most probably not due to Dark Matter particle annihilation. Future simultaneous observations with the present Cherenkov telescopes, the GLAST telescope and in the lower energies (X-rays, infrared and radio) will provide much better information on the source localization and variability of emission. This will shed new light on the nature of the high energy processes in the GC.

In order to disentangle the possible sources of the VHE $\gamma$-rays future observations should aim for:

- a more accurate determination of the position and extension of the VHE $\gamma$-ray source

- an extension of the spectrum of the VHE $\gamma$-rays towards higher and lower energies and a study of the variation of source position/extension with energy

- a search for time-variations in the VHE $\gamma$-ray flux and for simultaneous flux variability in infrared, X-rays and VHE $\gamma$-rays

- a further study of the connection between the diffuse $\gamma$-radiation along the Galactic Plane and the point-like VHE $\gamma$-ray source at the GC

- a search for neutrino and ultra high energy neutron emission from the GC.

## 4.2 Observation of VHE γ-Rays from HESS J1813-178

After a short introduction to the source HESS J1813-178 (section 4.2.1) the observational technique is discussed in section 4.2.2 and the procedure implemented for the data analysis is presented in section 4.2.3. Finally, this observation is put in the perspective of multifrequency data in section 4.2.4 and the results are summarized in section 4.2.5. The results of this thesis as presented in this section have been published in Albert et al. (2006a).

### 4.2.1 Introduction

In the Galactic Plane scan performed with the HESS Cherenkov telescopes array in 2004, with a flux sensitivity of 3% of the Crab Nebula flux for γ-rays above 200 GeV, eight sources were discovered (Aharonian et al. 2005a, 2006a), see also section 1.5.1. One of the newly detected γ-ray sources was HESS J1813-178. At the beginning, HESS J1813-178 could not be identified and was assumed to be a "dark particle accelerator," without reported counter-parts at lower frequencies.

Since the original discovery, HESS J1813-178 has been associated with the supernova remnant SNR G12.82-0.02 (Ubertini et al. 2005; Brogan et al. 2005; Helfand et al. 2005). One may still not exclude this coincidence being the result of just a chance association. Aharonian et al. (2005a) state a probability of 6% that one of their new sources is by chance spatially consistent with an SNR. Nevertheless, the properties of SNR G12.82-0.02, the multifrequency spectral energy distribution, and the flux and spectrum of the VHE γ-rays detected from this direction appear to be consistent with an SNR origin.

HESS J1813-178 has been found to be nearly point-like ($\sigma_{\text{source}} = 2.2'$) by Aharonian et al. (2006a). Given the size of the SNR, the angular resolution of the HESS telescope, and the depth of the observations, the source size does not rule out a possible shell origin. The γ-ray source lies at $10'$ from the center of the radio source W33. This patch of the sky is highly obscured and has indications of being a recent star formation region (Churchwell 1990).

### 4.2.2 Observations with the MAGIC Telescope

At La Palma, HESS J1813-178 culminates at about 47° zenith angle (ZA). The large ZA implies a high energy threshold of about 400 GeV for MAGIC observations. It also provides a large effective collection area, see section 3.6.5. The sky region around the location of HESS J1813-178 has a relatively high and non-uniform level of light, as shown in figure 4.7. Within a distance of 1° from HESS J1813-178, there are no stars brighter than $8^{\text{th}}$ magnitude, with the star field being in general brighter in the region south west of the source.

The MAGIC observations were carried out in the false-source tracking (wobble) mode, see section 2.3. The sky directions (W1, W2) to be tracked, are such that in the camera the sky field relative to the source position is similar for both wobble spots, see figure 4.7. The source direction is in both cases 0.4° offset from the camera center. In figure 4.8 these two

138                    4. Observation of VHE γ-Rays from Galactic Sources

tracking positions are shown by white stars. During wobble mode data taking, 50% of the data is taken at W1 and 50% at W2, switching (wobbling) between the 2 directions every 30 minutes. This observation mode allowed a reliable background estimation, although the observations had to be conducted at a relatively large ZA and the star field around the source is inhomogeneous.

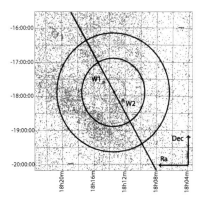

Figure 4.7: *The star field around HESS J1813-178. Stars up to a magnitude of 14 are shown. The two big circles correspond to distances of 1° and 1.75° from HESS J1813-178, respectively. The wobble positions W1 and W2 are given by the full circles. The full black line represents the Galactic Plane.*

### 4.2.3 Data Analysis

HESS J1813-178 was observed for a total of 25 hours in the period June-July 2005 (ZA ≤ 52°). In total, about 15 million triggers were recorded. The data were calibrated and analyzed using the procedure outlined in chapter 3: The Cherenkov signal charge and arrival time were reconstructed in FADC counts using the digital filter (see section 3.1). After calibration of the data (see section 3.2.1), image cleaning tail cuts were applied (see section 3.2.3) at levels of 10 and 5 photoelectrons for core and boundary pixels, respectively. The camera images are parameterized, see section 3.2.4. After the image cleaning and rejection of a few runs, which were affected by hardware problems, about 10 million events remained for further analysis.

In this analysis, the Random Forest method (see section 3.2.5) was applied for the γ/hadron separation and the energy estimation. For the training of the Random Forest a sample of Monte Carlo (MC) generated γ-ray showers was used to represent the γ-rays and a randomly chosen sub-set of the measured data was used to represent the background. The MC γ-ray showers were generated with a ZA between 47° and 54° and with energies between 10 GeV and 30 TeV. The spectral index of the generated differential spectrum $dN_\gamma/dE \sim$

## 4.2 Observation of VHE γ-Rays from HESS J1813-178

$E^\Gamma$ was chosen as $\Gamma = -2.6$, in agreement with the MAGIC-observed energy spectrum of the Crab Nebula at high energies, see section 3.4.3. The source-position independent image parameters Size, Width, Length and Concentration, as well as the source-position dependent parameter Dist and the third moment along the major image axis, were selected to parameterize the shower images (see section 3.2.4). After the training, the Random Forest method allows to calculate for every event a parameter, dubbed hadronness, which is a measure of the probability that the event belongs to the background. The γ-ray sample is defined by selecting showers with a hadronness below a specified value. An independent sample of MC γ-ray showers was used to determine the efficiency of the applied cuts.

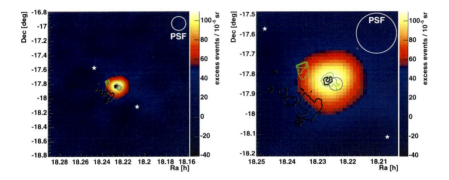

Figure 4.8: *Sky maps of γ-ray candidate excess events (background subtracted) from a sky region around HESS J1813-178 for an image Size ≥ 600 photoelectrons (corresponding to an energy threshold of about 1 TeV). The right sky map shows a magnified view around the source of the left sky map. A Gaussian folding has been applied, see text. Overlayed are contours of 90 cm VLA radio (black) and ASCA X-ray data (green) from (Brogan et al. 2005). The violet circle shows the position of the INTEGRAL source (Ubertini et al. 2005). The two white stars denote the tracking positions W1, W2 in the wobble mode.*

In order to develop and verify the analysis at high zenith angles (see section 3.6.5), Crab data in the interesting ZA range around 50° were taken in January 2005. From that sample the Crab energy spectrum was determined. The flux level as well as the spectral slope were found to be consistent within 1 standard deviation statistical uncertainty with other existing measurements (Hillas et al. 1998; Aharonian et al. 2004a) as well as the Crab spectrum measured with MAGIC at low zenith angles, see also section 3.6.5.

For each event, its original sky position is reconstructed by using the Disp-method, see section 3.2.7. At this stage only source independent image parameters are used in the RF training. Figure 4.8 shows the sky map of γ-ray candidate events (background subtracted, see section 3.3.2) from a sky region around HESS J1813-178. The sky map is folded (as described in section 3.3.2) with a two-dimensional Gaussian with a standard deviation of 0.072° and a maximum of one (the MAGIC γ-ray PSF is ~ 0.1°, see section

# 4. Observation of VHE γ-Rays from Galactic Sources

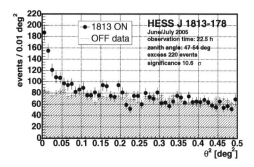

Figure 4.9: *Distributions of $\theta^2$ values with respect to the measured source position and three background control regions, see text, for Size $\geq$ 600 photoelectrons (corresponding to an energy threshold of about 1 TeV).*

3.6.2). To provide a good angular resolution a tight hadronness cut and a lower Size cut of 600 photoelectrons have been applied. The Size cut corresponds to an energy threshold of about 1 TeV. The sky map is overlayed with contours of 90 cm VLA radio and ASCA X-ray data from Brogan et al. (2005). The violet circle shows the position of the INTEGRAL source (Ubertini et al. 2005). The VHE γ-ray excess is centered at (RA, Dec)=($18^h13^m27^s$, $-17°48'40''$). The error of the excess location is dominated by the systematic pointing uncertainty of the MAGIC telescope of 2' (for a description of the MAGIC telescope drive system see section 2.2.2). In future it may be further reduced with an absolute calibration of the MAGIC star field monitor, see section 3.2.7. The observed excess position coincides well with the position of SNR G12.82-0.02 and the position of the VHE γ-ray source observed by Aharonian et al. (2006a) (RA,Dec)=($18^h13^m38^s$, $-17°50'33''$), who cite a statistical error of 18" but do not give a systematic error. Apart from the main excess coincident with HESS J1813-178 there are no other significant excesses present.

Figure 4.9 shows the distributions of the squared angular distance, $\theta^2$, between the reconstructed shower direction and the measured source position (black points) as well as the normalized distribution of the $\theta^2$ values with respect to three background control regions, see section 3.5.2. The observed excess in the direction of HESS J1813-178 has a significance of 10.6 standard deviations ($\theta^2 \leq 0.05 \deg^2$), according to equation 17 of Li&Ma (1983). Within errors the source position and the flux level are consistent with the measurement of HESS (Aharonian et al. 2006a).

For the determination of the energy spectrum, the RF was trained including the source dependent image parameters Dist and third moment of the photoelectron distribution with respect to the nominal excess position. A loose cut on the hadronness was used: Above the low energy turn-on, the cut efficiency reaches about 70% corresponding to an effective collection area for γ-ray showers of about 180000 m$^2$. Figure 4.10 shows the reconstructed very high energy γ-ray spectrum of HESS J1813-178 after the unfolding of the instrumental energy resolution (see section 3.4.3). The horizontal bars indicate the bin size in energy,

## 4.2 Observation of VHE γ-Rays from HESS J1813-178

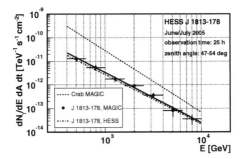

Figure 4.10: *Reconstructed VHE γ-ray spectrum of HESS J1813-178. The spectral index is $-2.1 \pm 0.2$ and the integral flux above 400 GeV is about 8% of the Crab Nebula (statistical errors only). The dashed line shows the spectrum of the Crab Nebula as measured by MAGIC, see section 3.4.3. The dot-dashed line shows the results of the HESS collaboration (Aharonian et al. 2006a).*

the marker is placed in the bin center on a logarithmic scale. The full line shows the result of a forward unfolding procedure as described in section 3.4.3: A simple power law spectrum is assumed for the true differential γ-ray flux. The parameters of the power law are determined by fitting the predicted differential flux to the measured energy spectrum. The result of the fit is given by ($\chi^2$/n.d.f = 7.3/7):

$$\frac{dN_\gamma}{dA dt dE} = (3.3 \pm 0.5) \times 10^{-12} (E/\text{TeV})^{-2.1 \pm 0.2} \text{ cm}^{-2}\text{s}^{-1}\text{TeV}^{-1}.$$

The quoted errors (1$\sigma$) are purely statistical. The systematic error is estimated to be 30% in the integral flux level and 0.2 in the spectral index, see section 3.7. Within errors the flux is steady at the timescales explored within these observations (weeks), as well as in the year-long time-span between the MAGIC and HESS observations.

### 4.2.4 Multiwavelength Source Modeling

Shortly after the discovery of HESS J1813-178, X-ray emission was found in ASCA data coming from the source AX J1813-178 (also known as AGPS 273.4-17.8) by Brogan et al. (2005) and also by Ubertini et al. (2005). The X-ray emission detected by ASCA is predominantly non-thermal, and compatible with that expected from a PWN or SNRs. Pulsed X-ray emission has not been detected, but the quality and amount of the data does not allow to set strong constraints (Brogan et al. 2005). Statistically, the X-ray data are not good enough to unambiguously separate a pure power law energy spectrum from a power law plus a thermal contribution either. However, none of the two component fits yielded a significantly different absorbing column density or photon index when compared with a single power law fit (Brogan et al. 2005). The analysis of data from the INTEGRAL satellite also showed a luminous source at approximately the same location, in the 20-100 keV range (Ubertini et al. 2005). Recently, the ASCA observations were confirmed by the Swift satellite (Landi et al. 2006).

Simultaneously with this X-ray match, HESS J1813-178 was also found as a non-thermal source in radio data, using observations with the Very Large Array (VLA) of radio telescopes (Brogan et al. 2005; Helfand et al. 2005) (90 and 20 cm) and the Bonn (11 cm) (Reich et al. 1990), Parkes (6 cm)(Haynes et al. 1978), and Nobeyama (3 cm) (Handa et al. 1987) telescopes. These groups discovered a shell-type supernova remnant (SNR G12.8-0.0) spatially coinciding with HESS J1813-178 and the ASCA X-ray source. The spectral index of the differential radio spectrum $\nu dN_\gamma/d\nu \sim \nu^{-\alpha}$ was found to be $\alpha = 0.48 \pm 0.03$. The SNR G12.8-0.0 was not detected at a radio wavelength of 4 m (Brogan et al. 2005). There are no known radio pulsars detected at the HESS J1813-178 position (Manchester et al. 2005) to a limiting 1.4 GHz flux density of 0.2 mJy. Deeper searches may however reveal pulsars at lower fluxes, see e.g. (Camilo et. al. 2002). According to the 20 cm radio map of Helfand et al. (2005) the SNR G12.8-0.0 had an apparent diameter of about 2.5'. At a distance of 4 kpc this corresponds to a radius of about 1.5 pc.

Brogan et al. (2005) suppose that the non-detection at 4 m wavelength is due to free-free absorption (thermal bremsstrahlung absorption, i.e. absorption of electromagnetic waves by free moving electrons). The most natural candidate for opacity is the W33 nebula, lying close to the line of sight. Thus, SNR G12.8-0.0 should lie at or beyond the distance of W33, $\sim 4$ kpc. Moreover, the high column density derived from ASCA data ($N_H \sim 10^{23}$ cm$^{-2}$, (Brogan et al. 2005)) as compared with the integrated Galactic H I column density along nearby lines of sight ($N_H \sim 10^{22}$ cm$^{-2}$, Dickey & Lockman (1990)) suggests a significant source of absorption in the foreground.

Figure 4.11 shows the multi-wavelength emission coming from the direction of HESS J1813-178, including the new MAGIC data at very high energies. In the following sections 4.2.4.1 and 4.2.4.2, hadronic (neutral pion decay), and leptonic (inverse Compton) models for the VHE $\gamma$-ray emission of the source are developed and compared to the data. These models are described in detail in section 1.2.

#### 4.2.4.1 Hadronic Models

In the framework of a hadronic model for the $\gamma$-ray emission (see section 1.2.1) one assumes that the observed VHE $\gamma$-ray flux is dominated by $\gamma$-rays from $\pi^0$ decay produced in collisions of accelerated VHE protons with ambient matter. In this model, the observed radio and X-ray emission would be due to a different population of high energy electrons. In case of a high magnetic field, as predicted by Völk et al. (2005), these electrons may only produce negligible amounts of IC $\gamma$-rays.

As a first step towards a hadronic emission model, the total energy of accelerated cosmic ray hadrons, which are stored at the source, is estimated from the observed $\gamma$-ray flux. In this model, for reasonable assumptions about the source distance and the ambient matter density the total energy of relativistic hadrons in about 10% of the average energy released by a supernova explosion. This makes a hadronic emission model tenable and a more detailed model for the emission spectrum is developed. It fits well to the VHE $\gamma$-ray data.

## 4.2 Observation of VHE γ-Rays from HESS J1813-178

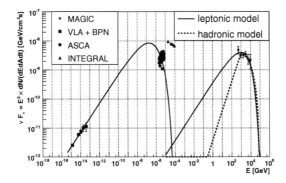

Figure 4.11: *Multi-wavelength emission coming from the direction of HESS J1813-178: Radio data are from the VLA, Bonn, Parkes, and Nobeyama observatories (Brogan et al. 2005), X-ray and hard X-ray data are from ASCA and INTEGRAL (Ubertini et al. 2005; Brogan et al. 2005). The lines show the developed leptonic and hadronic models for the HESS J1813-178 data. Details are given in the text.*

The observed γ-ray flux of HESS J1813-178 translates into an energy flux $w_\gamma$ between 400 GeV and 10 TeV of

$$w_\gamma(0.4-10 \text{ TeV}) = \int_{0.4 \text{ TeV}}^{10 \text{ TeV}} E \frac{dN_\gamma}{dAdtdE} dE \approx 1.0 \times 10^{-11} \frac{\text{TeV}}{\text{cm}^2\text{s}} . \quad (4.1)$$

The total γ-ray luminosity of the source at a distance $D$ in this energy range is then (assuming an isotropic γ-ray emission)

$$L_\gamma(0.4-10 \text{ TeV}) = 4\pi D^2 w_\gamma(0.4-10 \text{ TeV}) \approx 1.9 \times 10^{34} \frac{\text{TeV}}{\text{s}} \left(\frac{D}{4 \text{ kpc}}\right)^2 . \quad (4.2)$$

Let us in the following assume that the whole observed γ-ray flux is produced by proton proton interactions. This allows to determine the total energy in accelerated protons in the range 4 – 100 TeV required to provide the γ-ray luminosity, see equation 1.10 and 1.11:

$$W_p(4-100 \text{ TeV}) = t_{pp\rightarrow\gamma} \times L_\gamma(0.4-10 \text{ TeV}) = 1.2 \times 10^{50} \text{ TeV} \left(\frac{D}{4 \text{ kpc}}\right)^2 \left(\frac{n}{1 \text{ cm}^{-3}}\right)^{-1} . \quad (4.3)$$

Here it is assumed that the whole population of high energy cosmic ray protons interact with a target of constant density $n$ (assumed to only consist of protons).

According to section 1.2.1 the spectral index $\alpha$ of the hadrons is about equal to the index $\Gamma$ of the observed γ-rays. If the power-law proton spectrum (see equation 1.12) with differential spectral index $\alpha = \Gamma = 2.1$ continues down to $E_p \sim 1$ GeV and a maximum proton energy of 100 TeV is assumed, the total energy in protons is estimated to be

$$W_p(1\text{GeV} - 100\text{TeV}) = 10^{51} \text{ erg} \left(\frac{V_{\text{CR}}}{V_{\text{emission}}}\right) \left(\frac{D}{4 \text{ kpc}}\right)^2 \left(\frac{n}{1 \text{ cm}^{-3}}\right)^{-1} . \quad (4.4)$$

Thereby, $V_{\text{emission}}$ is the volume in which the accelerated high energy protons interact with the ambient target protons. In general it may be a subset of the total volume $V_{\text{CR}}$, which is filled with the accelerated cosmic rays. There are two cases for the relation between $V_{\text{emission}}$ and $V_{\text{CR}}$:

In the first case, $V_{\text{emission}} = V_{\text{CR}}$. A distance of 4 kpc and a matter density of $n = 10$ cm$^{-3}$ translate into an acceleration efficiency of 10% of the total SNR power of $10^{51}$ erg in accordance to theoretical models (see section 1.5.2). If the $\gamma$-ray production region is the whole SNR volume ($\sim 1.5$ pc radius at 4 kpc distance) this corresponds to a total target mass of about $V_{\text{emission}} \cdot n = 3 M_\odot$. In case the $\gamma$-rays are only produced in a shell of 15% thickness of the SNR radius, this would require a shell mass of about $V_{\text{emission}} \cdot n = 1 M_\odot$. These amounts of target material can be supplied by the ejecta of the SNR explosion.

In the second case $V_{\text{emission}} < V_{\text{CR}}$. The target mass for the high energy protons can be provided by a small cloud located in a region close to the SNR shell. As an example, a cloud of hydrogen molecules of 2 solar masses, with a density of 200 cm$^{-3}$, has less than 0.5 pc radius. This is compatible with the size of the SNR itself.

For reasonable parameters of the SNR and its environment the observed power of VHE $\gamma$-rays may be due to the acceleration of hadronic cosmic rays in the SNR with an efficiency of about 10%. Therefore, in the following the VHE $\gamma$-ray emission spectrum is modeled in more detail:

The hadronic model for the $\gamma$-ray emission shown in figure 4.11 was computed using the $\delta$-functional approximation, see section 1.2.1. The proton distribution is assumed to be (see equation 1.12):

$$\frac{dN_p(E_p)}{dV dE_p} = A_p (E_p/\text{GeV})^{-\alpha} \exp(-E/E_{\text{max}}) . \quad (4.5)$$

The spectral index was taken as equal with the spectral index of the reconstructed $\gamma$-ray spectrum $\alpha = 2.1$ and a cut-off energy of $E_{\text{max}} = 100$ TeV of the proton spectrum was assumed. The fit of this hadronic model to the MAGIC data yielded ($\chi^2/\text{n.d.f} = 2.1/5$) a proton density normalization of:

$$A_p = (2.2 \pm 0.2) \times 10^{-5} \times \left(\frac{V_{\text{SNR}}}{V_{\text{emission}}}\right) \left(\frac{D}{4 \text{ kpc}}\right)^2 \left(\frac{n}{1 \text{ cm}^{-3}}\right)^{-1} , \quad (4.6)$$

where $V_{\text{SNR}} \approx 4 \cdot 10^{56}$ cm$^3$ is the volume of the SNR. Higher cut-off energies of a few $10^{15}$ eV also give good fits to the data. The corresponding figure is omitted here. The source is thus compatible with proton acceleration up to the knee of the energy spectrum of around $10^{15}$ eV.

## 4.2 Observation of VHE γ-Rays from HESS J1813-178

The total energy of the accelerated protons is then given by:

$$W_p(E_p > 1\text{GeV}) = V_{\text{CR}} \int_{1\text{ GeV}}^{\infty} \frac{\mathrm{d}N_p(E_p)}{\mathrm{d}V\mathrm{d}E_p} E_p \mathrm{d}E_p \qquad (4.7)$$

$$= 10^{51} \text{erg} \left(\frac{V_{\text{CR}}}{V_{\text{emission}}}\right) \left(\frac{D}{4 \text{ kpc}}\right)^2 \left(\frac{n}{1 \text{ cm}^{-3}}\right)^{-1}, \qquad (4.8)$$

which is similar to the result obtained from the more coarse approximation above (equation 4.4).

To summarize, both the flux level as well as the spectral behavior of the VHE γ-radiation from HESS J1813-178 can be explained by the acceleration of hadronic cosmic rays in the SNR and their interactions with matter. The target matter may either be supplied by the ejecta of the SN explosion itself or by a small molecular cloud in the vicinity of the SNR. The observed radio and X-ray emission may be produced by an additional population of VHE electrons, which are accelerated in a similar manner as the protons (Ellison & Reynolds 1991). In case of a strong magnetic field only few electrons are needed for the radio and X-ray emission, such that they may only produce negligible amounts of VHE γ-rays.

### 4.2.4.2 Leptonic Models

In the case of a leptonic model for the VHE γ-ray emission (see section 1.2.2), it is assumed that the VHE γ-ray flux is produced by the inverse Compton up-scattering of low energy ambient photons by high energy electrons. Following Aharonian, Atoyan & Kifune (1997), the cosmic microwave background is the dominating source of target photons. Possible additional radiation fields like starlight and the synchrotron radiation produced by the same electron population which also produces the VHE γ-ray flux have been neglected.

The distribution of the relativistic electrons in the source is modeled as a power law spectrum with an exponential cut-off (see equation 1.20):

$$\mathrm{d}N_e/(\mathrm{d}V\mathrm{d}E_e) = A_e(E_e/\text{GeV})^{-\alpha_e} \exp\left(-E_e/E_{\text{max,e}}\right). \qquad (4.9)$$

Several different values of the parameters $A_e$, $\alpha_e$ and $E_{\text{max}}$ produce reasonable good fits to the measured spectrum of VHE γ-rays, e.g. $\alpha_e = 2.0$ and $E_{\text{max,e}} = 20$ TeV; and $\alpha_e = 2.1$ or 2.2 and $E_{\text{max,e}} = 30$ TeV. For all these models the total energy in relativistic electrons is at the percent level of the average supernova explosion energy of $10^{51}$ erg.

Furthermore, the leptonic model can be further constrained if one assumes that the radio and X-rays detected from SNR G12.8-0.0 are due to synchrotron radiation from the same electrons that produce the VHE γ-rays by the inverse Compton process. The spectral index of the differential radio spectrum $\nu \mathrm{d}N_\nu/\mathrm{d}\nu \sim \nu^{-\alpha_r}$ is $\alpha_r = 0.48 \pm 0.03$. It is linked to the spectral index $\alpha_e$ of the electron spectrum by $\alpha_e = 2\alpha_r + 1$ (see e.g. Aharonian (2004) for a review), yielding $\alpha_e = 1.96 \pm 0.06$.

For computing the synchrotron emission, one has an additional freedom, given by the unknown fraction $f_B$ of the inverse Compton emitting volume which is filled with magnetic fields. One obtains good fits to the radio data with magnetic fields between 5 and 10 μG and filling fractions between 15 and 30%. Both ranges are reasonable for SNR environments, see e.g. Jun & Jones (1999); Allen et al. (2001); Lazendic et al. (2004). Nevertheless, Völk et al. (2005) predict higher magnetic fields in SNR shocks, exceeding 100 $\mu$G. The spectral uncertainty in the X-ray regime (ASCA data) does not allow, however, to draw definitive conclusions upon the best fit values of the parameters involved, see also the discussion in (Brogan et al. 2005).

In the leptonic model shown in figure 4.11, a magnetic filling fraction of $f_B = 20\%$, a magnetic field of 10 $\mu$G, a maximum electron energy of $E_{\text{max},e} = 20$ TeV and $\alpha_e = 2.0$ have been adopted. The normalization of the electron spectrum is

$$A_e \approx 2 \times 10^{-7} \times \frac{V_{\text{SNR}}}{V_{\text{IC}}} \left(\frac{D}{4 \text{ kpc}}\right)^2. \qquad (4.10)$$

The total energy in relativistic electrons is then given by:

$$W_e(E_e > 1\text{MeV}) = V_{\text{IC}} \times \int_{1 \text{ MeV}}^{\infty} \frac{dN_e}{(dV dE_e)} E_e dE_e = 2.2 \times 10^{48} \text{erg} \times \left(\frac{D}{4 \text{ kpc}}\right)^2. \qquad (4.11)$$

This model is similar to one of the models presented by Brogan et al. (2005) (blue line in their figure 3, $E_{\text{max},e} = 30$ TeV, $\alpha_e = 2.0$), although they did not yet have a VHE $\gamma$-ray spectrum. The soft X-ray data (ASCA) plotted in figure 4.11 correspond to those shown by Brogan et al. (2005) (black data points in their figure 3). The presented leptonic model, as well as Brogan et al.'s, is in rough agreement with MAGIC and HESS data, the radio and roughly also with the ASCA X-ray data, within the uncertainty of the latter.

However, the hard X-ray data of INTEGRAL (Ubertini et al. 2005) cannot be fitted well with the same electron population that is assumed to produce the radio and X-ray (ASCA) data as synchrotron emission and the VHE $\gamma$-rays by IC upscattering of the CMB. In the framework of leptonic models, the high flux of hard X-rays from the INTEGRAL source must come from a different electron population, perhaps one generated by a compact object. If such an object exist, it might also contribute to the VHE $\gamma$-ray flux. As one can see in figure 4.8, the radio SNR shell matches spatially well with the main excess of the ASCA X-ray data. Contrary to that, the INTEGRAL source is offset from the radio SNR, but it coincides with the tail of the ASCA X-ray source. Further high-resolution multi-frequency observations in the X-ray and hard X-ray regime are needed to better constrain the electron populations in leptonic models.

To summarize, it is possible to explain the observed spectrum of VHE $\gamma$-rays by the IC upscattering of CMB photons by VHE electrons. The radio and ASCA X-ray data may be due to synchrotron radiation of the same electrons. Nevertheless, the INTEGRAL data cannot be explained in such a scenario. They may be due to a second electron population.

### 4.2.5 Concluding Remarks

The observation of HESS J1813-178 using the MAGIC Telescope confirms a new very high-energy $\gamma$-ray source in the Galactic Plane. A reasonably large data set was collected from observations at large zenith angles to infer the spectrum of this source up to energies of about 10 TeV. Between 400 GeV and 10 TeV the differential energy spectrum can be fitted

## 4.2 Observation of VHE $\gamma$-Rays from HESS J1813-178

with a power law of slope $\Gamma = -2.1 \pm 0.2$. These data can be used to cross-calibrate HESS and MAGIC, their independent observations show satisfactory agreement.

Multifrequency data in the radio, X-ray, and VHE $\gamma$-ray band imply a connection between HESS J1813-178 and SNR G12.82-0.02 (Helfand et al. 2005; Ubertini et al. 2005; Brogan et al. 2005). Generally, hard $\gamma$-ray spectra are expected from SNRs due to Fermi acceleration of cosmic rays, see e.g. Torres et al. (2003) for a review. The hard spectrum determined for HESS J1813-178 may be a further hint for its association with the SNR G12.82-0.02.

Present data are not sufficient to discriminate between existing models for different acceleration mechanisms. Future observations at lower energies with improved $\gamma$-ray telescopes and/or the GLAST satellite will undoubtedly permit to shed more light on the existing leptonic and hadronic models. Decisive information concerning hadronic acceleration mechanisms is also likely to come from future neutrino telescopes like IceCube (Ahrens et al. 2004) or KM3NeT (Katz 2006).

## 4.3 Observation of VHE γ-Rays from HESS J1834-087

In this section, after a short introduction to the source HESS J1834-087 (section 4.3.1), the observational technique used and the procedure implemented for the data analysis are presented in sections 4.3.2 and 4.3.3. Thereafter, the VHE γ-ray source is put in the perspective of the molecular environment found in the region of W41. $^{13}CO$ and $^{12}CO$ emission maps are presented, which show the existence of a massive molecular cloud in spatial superposition with HESS J1834-087. The results are discussed in the framework of leptonic and hadronic models for the VHE γ-ray emission in section 4.3.4. Finally, section 4.3.5 presents some conclusions. The results of this thesis as presented in this section were published in (Albert et al. 2006c).

### 4.3.1 Introduction

In the Galactic Plane scan performed with the HESS array of Cherenkov telescopes in 2004, with a flux sensitivity of 3% of the Crab Nebula flux for γ-rays above 200 GeV, eight sources were discovered (Aharonian et al. 2005a, 2006a). One of the newly detected γ-ray sources is HESS J1834-087 which is found to be, in projection, spatially coincident with SNR G23.3-0.3 (W41). The possibility of a random correlation between the VHE γ-ray source and SNR G23.3-0.3 was estimated to be 6% for the central region of the Galaxy (Aharonian et al. 2005a). The high energy source could also be connected to the "old" pulsar PSRJ1833-0827 (Gaensler & Johnston 1995), which would be energetic enough as to power HESS J1834-087. However, its location at 24 arc minutes away from the center of HESS J1834-087 renders an association unlikely (Aharonian et al. 2005a, 2006a). In addition, there is also no extended pulsar wind nebula (PWN) detected so far, whereas HESS J1834-087 was found to be extended: A brightness distribution $\rho \sim \exp(-r^2/2\sigma^2)$ with a size $\sigma = (0.09 \pm 0.02)°$ was reported by Aharonian et al. (2006a).

### 4.3.2 Observations with the MAGIC Telescope

At La Palma, HESS J1834-087 culminates at about 37° zenith angle (ZA). At this ZA the energy threshold for MAGIC observations is higher, but also, the effective collection area is larger (see section 3.6.5) than for observations at zenith. The sky region around the location of HESS J1834-087 has a relatively high and non-uniform level of background light, see figure 4.12. Within a distance of 1° from HESS J1834-087, there are 3 stars brighter than $8^{th}$ magnitude, with the star field being brighter in the region NW of the source. MAGIC observations were carried out in the false-source tracking (wobble) mode, see section 2.3. The sky directions (W1, W2) to be tracked are such that in the camera the sky brightness distribution relative to W1 is similar to the one relative to W2. The source direction is in both cases 0.4° offset from the camera center. These two tracking positions are shown by the full points in figure 4.12 and white stars in figure 4.13. For each tracking position two background control regions are used, which are located symmetrically to the source region (denoted by the central white circle) with respect to the camera center. During

## 4.3 Observation of VHE γ-Rays from HESS J1834-087

wobble mode data taking, 50% of the data is taken at W1 and 50% at W2, switching (wobbling) between the 2 directions every 30 minutes. This observation mode allowed a reliable background estimation, although the observations had to be conducted at a medium-scale ZA and the star field around the source is inhomogeneous.

HESS J1834-087 was observed for a total of 20 hours in the period August-September 2005 (ZA ≤ 45°). In total, about 12 million triggers have been recorded.

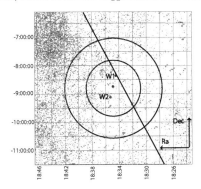

Figure 4.12: *The star field around HESS J1834-087: Stars up to a magnitude of 14 are shown. The two big circles correspond to distances of 1° and 1.75° from HESS J1834-087, respectively. The wobble positions W1 and W2 are given by the full points. The full black line represents the Galactic Plane.*

### 4.3.3 Data Analysis

The data analysis was carried out using the standard MAGIC analysis and reconstruction software (Bretz & Wagner 2003), the first step of which involves the calibration of the raw data (Gaug et al. 2005). It follows the general steps presented in chapter 3: After calibration, image cleaning tail cuts of 10 photoelectrons for image core pixels and 5 photoelectrons (boundary pixels) have been applied. These tail cuts are accordingly scaled for the larger size of the outer pixels of the MAGIC camera. The camera images are parameterized by so-called Hillas parameters (see section 3.2.4).

In this analysis, the Random Forest method (see section 3.2.5) was applied for the γ/hadron separation and the energy estimation (for a review about γ/hadron separation and energy estimation see e.g. Fegan (1997)). For the training of the Random Forest a sample of Monte Carlo (MC) generated γ-ray showers (Majumdar et al. 2005) was used to represent the γ-ray showers together with about 1% randomly selected events from the measured wobble data to represent the background. The MC γ-ray showers were generated between 35° and 45° ZA with energies between 10 GeV and 30 TeV with a Size distribution equal to the one of the selected data events for the training. The source-position independent image parameters Size, Width, Length, Concentration (Hillas 1985) and the absolute value of the third moment of the photoelectrons distribution along the major image axis were selected to parameterize the shower images. After the training, the Random Forest method allows to calculate for every event a parameter, the so-called hadronness, which is a measure of the probability that the event is not $\gamma$-ray like. The $\gamma$-ray sample is defined by selecting showers with a hadronness below a specified value. An independent sample of MC $\gamma$-ray showers was used to determine the cut efficiency.

The analysis at similar ZA was developed and verified using Crab Nebula data taken in September 2005, see also (Albert et al. 2006c). The determined Crab energy spectrum (see also section 3.4.3 and the dot-dashed line in figure 4.16) is consistent within one standard deviation statistical uncertainty with measurements from other experiments.

For each event the arrival direction of the primary $\gamma$-ray candidate in sky coordinates is estimated using the Disp-method (see section 3.2.7). A conservative lower Size cut of 200 photoelectrons is applied to select a subset of events with superior angular resolution. The corresponding analysis energy threshold is about 250 GeV.

Figure 4.13 shows the sky map of $\gamma$-ray candidates (background subtracted, see section 3.2.7) from a sky region around HESS J1834-087. The sky map was folded with a two-dimensional Gaussian with a standard deviation of 0.072° and a maximum of one. The MAGIC $\gamma$-ray PSF (standard deviation of a two dimensional Gaussian fit to the non-folded brightness profile of a point source, reconstructed with the Disp method) is 0.1° ± 0.01°, see section 3.6.2. The folding of the sky map serves to increase the signal-to-noise ratio by smoothing out statistical fluctuations. However, it somewhat degrades the spatial resolution to about 0.12° ± 0.01°. The sky map is overlayed with contours of 90 cm VLA radio data (green) from White et al. (2005) (20 cm radio data from the same reference are overlayed in the following figures) and $^{12}$CO emission contours from Dame et al. (2001) (black), integrated in the velocity range 70 to 85 km/s, the range that best defines the molecular cloud associated with W41.

4.3 Observation of VHE γ-Rays from HESS J1834-087        151

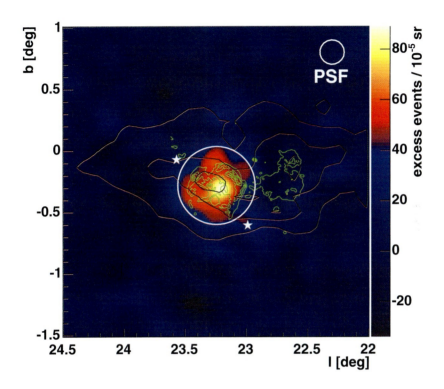

Figure 4.13: *Sky map of γ-ray candidate events (background subtracted) from a sky region around HESS J1834-087 for an energy threshold of about 250 GeV. A Gaussian folding was applied. The source is clearly extended with respect to the MAGIC PSF. The two white stars denote the tracking positions in the wobble mode. Overlayed are $^{12}CO$ emission contours (black) from Dame et al. (2001) and contours of 90 cm VLA radio data from White et al. (2005) (green). The $^{12}CO$ contours are at 25/50/75 K km/s, integrated from 70 to 85 km/s in velocity, the range that best defines the molecular cloud associated with W41. The contours of the radio emission are at 0.04/0.19/0.34/0.49/0.64/0.79 Jy/beam, chosen for best showing both SNRs G22.7-0.2 (right SNR shell in this sky map) and G23.3-0.3 (left SNR shell) at the same time. Clearly, the VHE γ-ray excess does not coincide with SNR G22.7-0.2. The central white circle (radius $\sqrt{0.1°}$) denotes the source region integrated for the spectral analysis. The center of this circle is at the measured position of the VHE γ-ray source.*

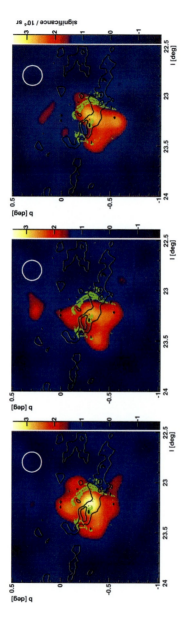

Figure 4.14: Morphology of HESS J1834-087 above three different lower cuts in Size (200, 300, 600 photoelectrons), corresponding to energy thresholds of 250, 360, 590 GeV. The color scale shows the excess significance. Overlayed are contours of 20 cm VLA radio data from White et al. (2005) (green) and $^{13}$CO emission contours (black) from Jackson et al. (2006). The contours of the radio emission are at 0.0095 Jy/beam. The $^{13}$CO contours are at 10/20/30 K km/s, integrated from 70 to 85 km/s in velocity, as was the $^{12}$CO data in figure 4.13. The white circle indicates the MAGIC PSF which is about 0.1 deg for all three lower Size cuts.

## 4.3 Observation of VHE γ-Rays from HESS J1834-087

The MAGIC excess is centered at (RA, Dec)=($18^h34^m27^s$, -8°42'40"). The statistical error is 0.5', the systematic pointing uncertainty is estimated to be 2' (see section 2.2.2). The large white circle in figure 4.13 is centered at the measured position of the VHE γ-ray source. This position is slightly offset from the assumed source position used to determine the two pointing directions in the wobble mode (indicated by the two white stars in figure 4.13). A fit of a two dimensional Gaussian brightness profile to the non-folded sky map yields, after subtraction of the MAGIC γ-ray PSF in quadrature, an intrinsic source extension of $\sigma = (0.14 \pm 0.04)°$, the extension reported by HESS is $(0.09 \pm 0.02)°$ (Aharonian et al. 2006a). Both, the center of gravity position and the extension of the excess, coincide well with the shell-type SNR G23.3-0.3. (W41).

Figure 4.14 shows images of HESS J1834-087 with three different lower cuts on Size (200, 300, 600 photoelectrons), corresponding to energy thresholds of about 250, 360 and 590 GeV. As in figure 4.13, the background subtracted sky maps are folded with a two-dimensional Gaussian, but here the color scale shows directly the excess significance. The total observed excess significances for $\theta^2 \leq 0.1\text{deg}^2$ (corresponding to the sky region inside the central white circle of figure 4.13) are $8.6\sigma$, $7.8\sigma$ and $7.3\sigma$ for the three lower cuts on Size. Overlayed are contours of 20 cm VLA radio data from White et al. (2005) (green) and $^{13}$CO emission contours (black) from Jackson et al. (2006). The contours of the radio emission are at 0.0035 Jy/beam. The $^{13}$CO contours are integrated from 70 to 85 km/s in velocity, as was the $^{12}$CO data in figure 4.13. For all three Size cuts the MAGIC PSF is about 0.1°, and the source position, extension and morphology stay roughly constant. The characteristics of the MAGIC observation are compatible within errors with the measurement of HESS (Aharonian et al. 2006a).

Figure 4.15 shows the distribution of the squared angular distance, $\theta^2$, between the reconstructed shower direction and the excess center as well as the $\theta^2$-distribution for the background, see sections 3.3.2 and 3.5.2. The observed excess in the direction of HESS J1834-087 has a significance of $8.6\sigma$ for $\theta^2 \leq 0.1\text{deg}^2$. The VHE γ-ray brightness distribution of the source is clearly non-Gaussian, as can also be seen from figures 4.13 and 4.14. The source may be composed of a point-like core and an extended plateau region.

For the spectral analysis a sky region of maximum angular distance of $\theta^2 = 0.1\text{deg}^2$ around the excess center (indicated by the white circle in figure 4.13) has been integrated. Figure 4.16 shows the reconstructed very high energy γ-ray spectrum ($dN_\gamma/(dE_\gamma dAdt)$ vs. true $E_\gamma$) of HESS J1834-087 after correcting (unfolding) for the instrumental energy resolution (see section 3.4.3). The horizontal bars indicate the bin size in energy, the markers are placed in the bin centers on a logarithmic scale. The full line shows the result of a forward unfolding procedure: A simple power law spectrum is fitted to the measured spectrum ($dN_\gamma/(dE_\gamma dAdt)$ vs. estimated $E_\gamma$) taking the full instrumental energy migration (true $E_\gamma$ vs. estimated $E_\gamma$) into account. The result is ($\chi^2/\text{n.d.f} = 7.4/7$):

$$\frac{dN_\gamma}{dAdtdE} = (3.7 \pm 0.6) \times 10^{-12} \left(\frac{E}{\text{TeV}}\right)^{-2.5\pm 0.2} \text{cm}^{-2}\text{s}^{-1}\text{TeV}^{-1}. \quad (4.12)$$

The quoted errors are purely statistical. The systematic error is estimated to be 35% in the flux level determination and 0.2 in the spectral index, see section 3.7. Within the observation time (weeks) no flux variations exceeding the measurement errors have been

# 4. Observation of VHE γ-Rays from Galactic Sources

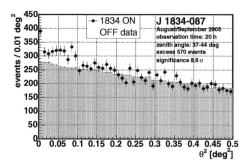

Figure 4.15: *Distributions of $\theta^2$ values for the source (full circles) and background control regions (shaded histogram), see text, for an energy threshold of about 250 GeV. It is evident that the source is clearly extended with respect to the MAGIC PSF ($\sigma^2 \sim 0.01 \deg^2$).*

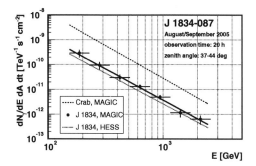

Figure 4.16: *VHE γ-ray spectrum of HESS J1834-087 (statistical errors only). The solid line shows the result of a power-law fit to the data points taking the full instrumental energy resolution into account. The dotted line shows the result of the HESS collaboration (Aharonian et al. 2006a). The dashed line shows the spectrum of the Crab Nebula as measured by MAGIC, see section 3.4.3.*

observed. The corresponding figure is not shown here. Also, the flux is compatible within errors with the measurement of HESS made one year earlier.

## 4.3.4 Discussion

SNRs as γ-ray sources were extensively discussed in the past (see section 1.5.2 and e.g. Torres et al. (2003) for a review). Due to the spatial coincidence between the VHE γ-ray

## 4.3 Observation of VHE $\gamma$-Rays from HESS J1834-087

source and the SNR G23.3-0.3 (W41), this SNR appears to be the natural candidate for generating the observed $\gamma$-ray emission. W41 is an asymmetric shell-type SNR with a hot spot in its center. It is included in Green's catalog (Green 2004) of SNRs, which states a diameter of 27' and a distance of about 4.8 kpc. The spectral index of the differential radio spectrum $\nu \mathrm{d}N_\nu/\mathrm{d}\nu \sim \nu^{-\alpha}$ is $\alpha = 0.5$. The flux density at 1 GHz is 70 Jy. W41 was mapped in radio with the VLA array at 330 MHz (Kassim 1992) and at 20 and 90 cm (Condon et al. 1998; White et al. 2005), following earlier studies (see e.g., Ariskin (1970), Shaver (1970), and references therein). It is partially overlapping with SNR G22.7-0.2 (see e.g. figure 9 of Kassim (1992)), however the latter is not in coincidence with the peak of the VHE $\gamma$-ray source (see figure 4.13 above).

Recently, Landi et al. (2006) observed the SNR G23.3-0.3 (W41) with the X-ray telescope (XRT) onboard the Swift satellite (6900 s exposure). They found three faint X-ray sources. All of them are too weak for a spectral analysis and only rough estimates of the flux can be given. One source is located close to the radio hot spot at the center of the SNR. Within the XRT positional uncertainty it may have an optical counter-part located at (RA, Dec)=($18^\mathrm{h}34^\mathrm{m}34.90^\mathrm{s}$, -8°44'49".6) listed in the USNO-B1.0 and 2MASS catalogues. The other two X-ray objects are found near the SNR shell.

Also the XMM X-ray satellite observed the sky region for about 20 ks. The data[2] confirm the X-ray sources observed by Landi et al. (2006) and indicate some additional weak point-like sources. No hard X-radiation spatially consistent with the W41 SNR was found in the INTEGRAL data (Bird et al. 2006).

In the following, the VHE $\gamma$-ray source is put in the perspective of the molecular environment found in the region of W41 in section 4.3.4.1. Thereafter, the possible production of the observed VHE $\gamma$-radiation by hadronic (section 4.3.4.2) or leptonic interactions (section 4.3.4.3) are discussed.

### 4.3.4.1 Molecular Clouds

Molecular hydrogen is very difficult to detect as it does not emit any prominent lines in the electromagnetic spectrum. Therefore, indirect methods using the detection of tracer molecules like CO are often applied, see section 1.5.5. In a large scale CO survey of the galaxy, W41 was associated with a very large molecular complex called "[23,78]" by Dame et al. (1986). There, it was concluded that there are probably two large clouds blended at that position in $l - b - v$ space (galactic longitude, galactic latitude and relative velocity corresponding to the redshift of the CO emission line). One of these clouds is located in the near side of the 4 kpc arm and another in the far side of the Scutum Arm (for an overview of the Milky Way see section 1.5.1). This conclusion was also earlier supported by a previous radio recombination line study by Gordon & Gordon (1970). The giant molecular cloud associated with W41 is best defined by integrating the CO emission from 70 to 85 km/s in velocity (see section 1.5.5). The CO emission peaks near l=23.3°, b=−0.3°, v=78 km/s; the near kinematic distance of this peak is 4.9 kpc. The peak is marked by the central black contour in figure 4.13, which lies very close to the VHE $\gamma$-ray source. The total $H_2$ mass of the cloud, integrated over the range l=22° to 24.25°, b=−0.75° to 0.5°, and v=70

---

[2]Publicly available at http://xmm.esac.esa.int/external/xmm_data_acc/xsa/index.shtml

to 85 km/s, and assuming a distance of 4.9 kpc, is $2.1 \times 10^6$ $M_\odot$. This mass is necessarily an upper limit since, as mentioned above, there is certainly an emission contribution from unrelated gas at the far kinematic distance. Still, the CO peak is so strong and well defined that it most likely arises from gas primarily at one location, near the VHE $\gamma$-ray source, rather than being a random blend of emissions from the near and far distances. The total $H_2$ mass of the CO emission peak in figure 4.13 (computed for the region l=23.2° to 23.4°, b=−0.35° to −0.15°, v=70 to 85 km/s) is $8.8 \times 10^4$ $M_\odot$. The higher-resolution $^{13}$CO map in figure 4.14, which was derived from the recently completed Galactic Ring survey (Jackson et al. 2006), confirms that the VHE $\gamma$-ray source lies towards the densest region of the giant molecular cloud. This region has a column density of about 0.06 g/cm². This value is small compared to the specific radiation length in gaseous hydrogen of 61.28 g/cm² (Yao et al. 2006), see also section 2.1.1.2. Therefore, even in the case that the molecular cloud is located between the observer and the $\gamma$-ray source, the $\gamma$-ray absorption in the cloud is negligible.

### 4.3.4.2 Hadronic Models

In the framework of a hadronic model for the $\gamma$-ray emission (see section 1.2.1), one assumes that the observed VHE $\gamma$-ray flux is dominated by $\gamma$-rays from $\pi^0$ decay produced in collisions of accelerated VHE protons with ambient matter. The observed radio and X-ray emission would, in this model, be due to a different population of high energy electrons. In case of a high magnetic field, as predicted by Völk et al. (2005), these electrons may only produce negligible amounts of IC $\gamma$-rays.

The luminosity of HESS J1834-087 is similar to the luminosity of HESS J1813-178 (see section 4.2), if the latter source is considered to be associated with SNR G12.8-0.0 at a distance of $\sim$ 4 kpc. One has to note, though, that HESS J1813-178 was found to be nearly point-like, whereas in the present case, a significant extension is observed. The strongest $\gamma$-ray emission comes from inside the SNR shell, in contrast to the well-studied shell-type SNRs RXJ1713.7-3946 and RXJ0852.0-4622 (Vela Junior) (Enomoto et al. 2002; Aharonian et al 2004c; Katagiri et al. 2005; Aharonian et al. 2005c; Enomoto et al. 2006). Also the $\gamma$-ray spectrum of HESS J1834-087 is steeper then that expected from a spectral index of about 2.1 for diffusive shock acceleration of hadronic cosmic rays (see sections 1.3 and 1.5.2).

However, these properties may be explained by different regions, where the cosmic ray protons are accelerated and where the VHE $\gamma$-rays are produced: The proton acceleration may proceed in the shock front in the SNR shell. The accelerated cosmic rays may than diffuse to the densest part of the molecular cloud (in projection inside the SNR shell). An energy dependent diffusion coefficient may cause a steepening of the cosmic ray spectrum and thus also produce a $\gamma$-ray spectrum steeper than 2.1.

The exact modelling of the cosmic ray acceleration and diffusion in the molecular cloud are beyond the scope of this thesis. In order to judge whether such a model is possible at all, the total energy of accelerated and stored cosmic ray hadrons is estimated from the observed $\gamma$-ray flux and compared with the average energy released by a supernova explosion:

The observed $\gamma$-ray flux of HESS J1813-178 translates into an energy flux between

## 4.3 Observation of VHE $\gamma$-Rays from HESS J1834-087

250 GeV and 2.5 TeV of

$$w_\gamma(0.25-2.5 \text{ TeV}) = \int_{0.25 \text{ TeV}}^{2.5 \text{ TeV}} E \frac{dN_\gamma}{dA dt dE} dE \approx 1.0 \times 10^{-11} \frac{\text{TeV}}{\text{cm}^2 \text{s}}. \quad (4.13)$$

The total $\gamma$-ray luminosity of HESS J1834-087 in this energy range at a distance $D$ is (assuming an isotropic $\gamma$-ray emission):

$$L_\gamma(0.25-2.5 \text{ TeV}) = 4\pi D^2 w_\gamma(0.25-2.5 \text{ TeV}) \approx 3.0 \times 10^{34} \frac{\text{TeV}}{\text{s}} \left(\frac{D}{5 \text{ kpc}}\right)^2. \quad (4.14)$$

If one assumes that the whole $\gamma$-ray flux is produced by hadronic interactions of high energy protons with ambient protons of uniform density $n$, this allows to determine the total energy in accelerated protons in the range $2.5 - 25$ TeV, required to provide the $\gamma$-ray luminosity, see equation 1.10 and 1.11:

$$W_p(2.5-25 \text{ TeV}) = t_{pp\to\gamma} \times L_\gamma(0.25-2.5 \text{ TeV}) = 1.8 \times 10^{50} \text{ TeV} \left(\frac{D}{5 \text{ kpc}}\right)^2 \left(\frac{n}{1 \text{cm}^{-3}}\right)^{-1}. \quad (4.15)$$

According to section 1.2.1 the spectral index $\alpha$ of the hadrons is about equal to the index $\Gamma$ of the observed $\gamma$-rays. If the power-law proton spectrum (see equation 1.12) with differential spectral index $\alpha = \Gamma = 2.5$ continues down to $E_p \sim 1$ GeV and has a maximum proton energy of 100 TeV, the total energy in protons is estimated to be

$$W_p(1 \text{ GeV} - 100 \text{ TeV}) = 2.1 \cdot 10^{52} \text{ erg} \left(\frac{V_{\text{CR}}}{V_{\text{emission}}}\right) \left(\frac{D}{5 \text{ kpc}}\right)^2 \left(\frac{n}{1 \text{cm}^{-3}}\right)^{-1}. \quad (4.16)$$

Thereby $V_{\text{emission}}$ is the volume in which the accelerated high energy protons interact with the ambient target protons. In general it may be a subset of the total volume $V_{\text{CR}}$, which is filled with the accelerated cosmic rays.

Assuming an acceleration efficiency of hadrons of 10% of the total supernova power of $10^{51}$ erg (see section 1.5.2), the required density $n$ of matter in the $\gamma$-ray production region is $\sim 20$ protons cm$^{-3}$ $V_{\text{CR}}/V_{\text{emission}}$.

Given the extension of HESS J1834-087, and the gas mass found in the innermost contour of the CO map, i.e. in close superposition with the VHE $\gamma$-ray source, there is enough mass to generate the VHE $\gamma$-radiation hadronically, even if only part of the gas is interacting with the SNR shock.

### 4.3.4.3 Leptonic Models

As in the case of HESS J1813-178, also the VHE $\gamma$-radiation from HESS J1834-087 may be due to the inverse Compton upscattering of low energy radiation fields (dominantly the CMB) by very high energy electrons. In this case the observed radio and X-radiation may be due to synchrotron radiation of the same electrons.

Nevertheless, there seems to be a spatial anti-correlation between the radio data and the VHE $\gamma$-rays. The VHE $\gamma$-ray source is brightest inside the SNR and not in the shell region. However, there may be an electron population inside the SNR, which produces the radio hot-spot, one of the X-ray sources and the VHE $\gamma$-ray.

The present data (especially in X-rays) are not sufficient to accurately model the multi-wavelength emission.

### 4.3.5 Conclusions

In summary, the observation of HESS J1834-087 using the MAGIC Telescope confirms a new extended VHE $\gamma$-ray source in the Galactic Plane. A reasonably large data set was collected from observations at medium-scale zenith angles to infer the spectrum of this source up to energies of a few TeV. Above 200 GeV, the differential energy spectrum can be fitted with a power law of slope $\Gamma = -2.5 \pm 0.2$. The results of the independent observations of the HESS and MAGIC telescopes are in agreement within errors concerning the level of flux, the spectral shape, the morphology, and the extension of the source. The coincidence of the VHE $\gamma$-ray source with SNR G23.3-0.3 (W41) poses this SNR as a natural counter-part, and although a massive molecular cloud has been identified in the region, the mechanism responsible for the VHE $\gamma$-radiation remains yet to be clarified.

# Chapter 5

# The Data Acquisition System Upgrade of MAGIC

Ground-based Atmospheric Air Cherenkov Telescopes (ACTs) are successfully used to observe very high energy (VHE) $\gamma$-rays from celestial objects. The light of the night sky (LONS) is a strong background for these telescopes. The $\gamma$-ray pulses being very short, an ultra-fast read-out of an ACT can minimize the influence of the LONS. This allows one to lower the so-called tail cuts of the shower image (see sectio 3.2.3) and the analysis energy threshold. Moreover, an ultra-fast read-out system allows to use timing information to separate $\gamma$-ray showers from backgrounds.

Fast "flash" analog-to-digital converters (FADCs) with GSamples/s are available commercially; they are, however, very expensive and power consuming. Here we present a novel technique of Fiber-Optic Multiplexing which uses a single 2 GSamples/s FADC to digitize 16 read-out channels consecutively. The analog signals are delayed by using optical fibers. The multiplexed (MUX) FADC read-out reduces the cost by about 85% compared to using one ultra-fast FADC per read-out channel.

Two prototype multiplexers, each digitizing data from 16 channels, were built and tested. In this chapter the ultra-fast read-out system will be described and the test results will be reported. The new system will be implemented for the read-out of the 17 m diameter MAGIC telescope camera.

In section 5.1 (see also section 2.2.6) the MAGIC experiment is briefly described in the context of the data acquisition (DAQ) system using ultra-fast FADCs. The specifications of the ultra-fast read-out are described in section 5.2, followed by the measured performance for the MUX-FADC prototype in laboratory tests (section 5.3) and as read-out of the MAGIC telescope (section 5.4). Section 5.5 is dedicated to discussions and conclusions about the prototype test of the MUX-FADC system. This work was published (Bartko et al. 2005b). Finally, section 5.6 summarizes the production of the full MUX-FADC system and the installation in the MAGIC telescope.

# 5. The Data Acquisition System Upgrade of MAGIC

## 5.1 DAQ System Upgrade Considerations

The camera of the MAGIC Telescope consists of 576 photomultiplier tubes (PMTs), which deliver about 2 ns FWHM fast pulses to the experimental control house, see section 2.2.4. The currently used read-out system (see section 2.2.6) is relatively slow (300 MSamples/s). To record the pulse shape in detail, an artificial pulse stretching to about 6.5 ns FWHM is used. This eliminates the characteristic signal distribution of an isolated muon, a $\gamma$-ray shower and a hadronic shower and causes more light of the night sky (LONS) to be integrated, which acts as additional noise. Thus the analysis energy threshold of the telescope is limited, and the selection efficiency of the $\gamma$-ray signal from different backgrounds is reduced.

For the fast Cherenkov pulses (2 ns FWHM), an FADC with 2 GSamples/s can provide at least four sampling points. This permits a reasonable reconstruction of the pulse shape and could yield an improved $\gamma$/hadron separation based on timing. Such an ultra-fast read-out can strongly improve the performance of MAGIC. The improved sensitivity and the lower analysis energy threshold will considerably extend the observation range of MAGIC, and allow one to search for very weak sources at high redshifts.

A few FADC products with $\geq$ 2 GSamples/s and a bandwidth $\geq$ 500 MHz are available commercially; they are, however, very expensive and power-consuming. To reduce the cost of an ultra-fast read-out system, a 2 GSamples/s read-out system has been developed at the Max-Planck-Institut für Physik in Munich. It uses the novel technique of Fiber-Optic Multiplexing, an approach possible because the signal duration (few ns) and the trigger frequency (typically $\sim$1 kHz) result in a very low duty cycle for the digitizer. The new technique uses a single FADC of 700 MHz bandwidth and of 2 GSamples/s to digitize 16 read-out channels consecutively. Groups of 16 analog signals are delayed by using optical fibers of 16 different lengths. A trigger signal is generated using a fraction of the light, which is branched off by fiber-optic light splitters before the delay fibers. With the Fiber-Optic Multiplexing a cost reduction of about 85% is achieved compared to using one FADC per read-out channel.

The suggested 2 GSamples/s multiplexed (MUX) FADC system will have a 10 bit amplitude resolution. For large signals the arrival time of the Cherenkov pulse can be determined with a resolution better than 200 ps. The system is relatively simple and reliable. All optical components and the FADCs are commercially available, while the multiplexer electronics has been developed at the MPI in Munich. Two prototype multiplexers, for 32 channels in total, were built and tested in-situ as read-out of the MAGIC telescope in La Palma in August 2004.

Monte Carlo (MC) based simulations predict different time structures for $\gamma$-ray and hadron induced shower images as well as for images of single muons (Chitnis & Bhat 2001; Mirzoyan et al. 2006). As shown in figure 5.1, $\gamma$-ray and hadron initiated showers as well as isolated muons have different distributions of the RMS value of the arrival times. The timing information is therefore expected to improve the separation of $\gamma$-ray events from the background events.

Figure 5.2 shows the mean amplitude (a, c) and time (b, d) profiles for $\gamma$-ray (c, d) and hadron (a, b) induced air showers images on the camera plane of the MAGIC telescope. The impact parameter is fixed to 120 m and the initial $\gamma$-ray energy is set to 100 GeV, while

## 5.2 The Ultra-fast Fiber-Optic MUX-FADC Data Acquisition System

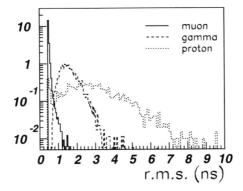

Figure 5.1: *Distributions of the RMS value of the arrival time distribution for all $\gamma$-ray, proton and isolated muon events measured by a 17m ultra-fast telescope. The events with RMS $\leq$ 0.7 ns are of muon origin. The RMS value of the arrival time distribution thus offers a separation power between muons and $\gamma$-ray induced showers. Figure from (Mirzoyan et al. 2006).*

the proton energy is set to 200 GeV (corresponding to about 1/4 of the Cherenkov light of a 100 GeV $\gamma$-ray shower). The profiles are obtained by averaging over many simulated showers (Mazin 2006). Although the total durations of $\gamma$-ray and hadron induced air showers are comparable, the photon arrival time varies smoothly over the $\gamma$-ray shower image while hadron shower images are structured in time. The time structure of the image can also provide essential information about the head and the tail of the shower.

### 5.2 The Ultra-fast Fiber-Optic MUX-FADC Data Acquisition System

The MAGIC collaboration is going to improve the performance of its telescope by installing a fast ($\geq$ 2 GSamples/s) FADC system, which fully exploits the intrinsic time structures of the Cherenkov light pulses. The requirements for such a system are the following:

- 10 bit resolution at a 2 GSamples/s sampling rate
- $\geq$ 500 MHz bandwidth of the electronics chain including the FADC
- up to 1 kHz sustained event trigger rate
- dead time $\leq$ 5%.

This section describes the specifications of the ultra-fast fiber-optic MUX-FADC data acquisition system: In section 5.2.1 the general multiplexing principle is explained. In

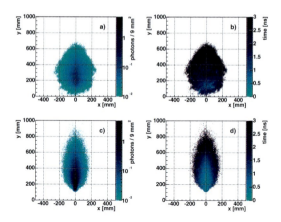

Figure 5.2: *Mean amplitude (a, c) and time (b, d) profiles for $\gamma$-ray (c, d) and hadron (a, b) induced air showers images on the MAGIC camera plane from MC simulations. The impact parameter is fixed to 120 m and the initial energy of the $\gamma$-ray is set to 100 GeV, while the proton energy is set to 200 GeV. The profiles are obtained by averaging over many simulated showers (Mazin 2006). The timing structure of the image can provide viable information about the head and the tail of the shower as well as help to discriminate between $\gamma$-ray and hadron induced showers.*

section 5.2.2 the optical splitters and delays are described and in section 5.2.3 details of the multiplexing electronics are presented. Finally, in section 5.2.4 the FADC read-out is specified.

## 5.2.1 General MUX-Principle

It is interesting to note that in experiments where FADCs are used to read out a multichannel detector in the common event trigger mode, only a tiny fraction of the FADC memory depth is occupied by the signal while the rest is effectively "empty" (Mirzoyan et al. 2002, 2001). One can try to correct this "inefficiency of use" by "packing" the signals of many channels sequentially in time into a single FADC channel, i.e. by multiplexing.

Following this simple idea, a multiplexing system with fiber-optic delays has been developed for the MAGIC telescope. The block diagram is shown in figure 5.3. The ultrafast fiber-optic multiplexer consists of three main components:

- fiber-optic delays and splitters
- multiplexer electronics: fast switches and controllers
- ultra-fast FADCs.

## 5.2 The Ultra-fast Fiber-Optic MUX-FADC Data Acquisition System

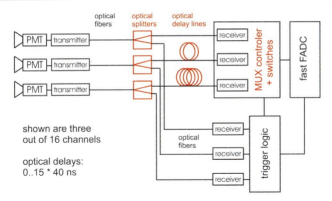

Figure 5.3: *Schematic diagram of the multiplexed fiber-optic ultra-fast FADC read-out. Part of the analog signal that arrives via the fiber-optic link from the PMT camera is branched off and fed into a majority trigger logic. The other part of the signal is consecutively delayed in optical fibers. Channels are connected consecutively one by one to the ultra-fast FADC, using fast switches. Thereby the noise from the other channels is efficiently blocked.*

After the analog optical link between the MAGIC PMT camera and the counting house the optical signals are split into two parts. One part of the split signal is used as an input to the trigger logic. The other part is used for FADC measurements after passing through a fiber-optic delay line of a channel-specific length.

The multiplexer electronics operates in the following way: The common trigger from the majority logic unit opens the switch of the first channel and allows the analog signal to pass through and be digitized by the FADC. All the other switches are closed during this time. When the digitization window for the first channel is over the corresponding switch is closed. The closed switch strongly attenuates the signal transmission by more than 60 dB for the fast MAGIC signals. Then the switch number two is opened such that the accordingly delayed analog signal from the second channel is digitized and so forth, one channel at a time until the last one is measured. In this way one "packs" signals from different channels in a time sequence which can be digitized by a single FADC channel.

Because of the finite rise and fall times of the gate signals for the switches and because of some pick-up noise from the switch one has to allow for some switching time between the digitization of two consecutive channels. The gating time for each channel was set to 40 ns, of which the first and last 5 ns are affected by the switching process.

For the use in MAGIC a $16 \rightarrow 1$ multiplexing ratio was chosen. 16 channels are read out by a single ultra-fast FADC channel. The chosen multiplexing ratio is a compromise between

- Dead time: the digitization of one event takes $16 * 40\,\text{ns} = 640$ ns. During this time no other event can be recorded. Compared to the maximum sustained trigger rate of up to 1 kHz this dead time is negligible.

- Noise due to cross-talk through the closed switches: The attenuated noise of the other channels could influence the active signal channel.

- Cost of the FADCs.

- Mechanical constraints, e.g. board size, length of wires and fibers.

### 5.2.2 Optical Delays and Splitters

Optical fibers were chosen for the analog signal transmission between the PMT camera and the counting house because they are lightweight, compact, immune to electro-magnetic pick-up noise, have practically no temperature effect and provide no pulse dispersion and attenuation (Lorenz et al. 2001). The signal attenuation at 1 km fiber length is about 2.3 dB for the chosen 850 nm wavelength of the VCSELs. The analog signal transmission offers a dynamic range larger than 60 dB.

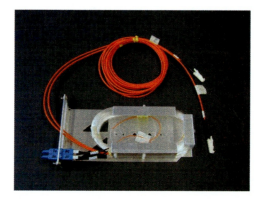

Figure 5.4: *Two channel fiber-optic delay module of 142 and 150 m length, corresponding to a delay of 710 and 750 ns, respectively. Mechanical dimensions: 235 mm * 130 mm (3U) * 35 mm (7 HP).*

Using fiber-optic delays ultra-fast analog signals can be delayed by several hundreds of ns. Thus a large number of successively delayed signals can be multiplexed and read out by a single channel FADC. Part of the analog signal has to be split off before the delay lines for trigger. Therefore, fiber-optic splitters of type $1 \rightarrow 2$ are used.

Figure 5.4 shows a module containing two optical delay lines of 142 m and 150 m length, corresponding to a delay of 710 ns and 750 ns. Figure 5.5 shows a module of four graded index (GRIN)-type fiber-optic splitters with 50:50 splitting ratio (for a technical description see Lipson & Harvey (1983)). The modules have standardized outer dimensions and can

## 5.2 The Ultra-fast Fiber-Optic MUX-FADC Data Acquisition System

Figure 5.5: *Four channel fiber-optic splitter module, GRIN technology and 50:50 splitting ratio. The outer dimensions are: 235 mm * 130 mm (3U) * 35 mm (7 HP).*

be assembled in 3U hight 19" crates. The splitters and optical delay lines are commercially available from the company Sachsenkabel[1].

### 5.2.3 MUX Electronics

The multiplexer electronics consist of four stages. The first stage is a fiber optic receiver, where the signals from the optical delay lines are converted back to electrical pulses using PIN photo diodes. In a second stage, part of the electrical signal is branched off and transferred to a monitor output. The third stage consists of ultra-fast switches which are activated one at a time. In the last stage all 16 channels are summed to one output. The multiplexed signals are then transferred via short 50 $\Omega$ coaxial cables to the FADC channels. Table 5.1 summarizes the specifications of the multiplexer electronics.

| | |
|---|---|
| Mechanical size | 370 mm (9 U) * 220 mm * 30 mm (6 HP) |
| Number of channels | 16 |
| Analog input | via 50/125 $\mu$m graded index fiber, E2000 connector |
| Gain | 25, including the VCSEL transmitter |
| Dynamic range | max output amplitude: 1 V |
| Power supplies | +12 V, ±5 V |
| Power dissipation | ~20 W |
| Trigger input | LVDS |

Table 5.1: *Specifications of the electronics for analog signal multiplexing.*

---

[1] http://www.sachsenkabel.de

# 5. The Data Acquisition System Upgrade of MAGIC

Figure 5.6: *Photo of the printed circuit board for analog signal multiplexing: It consists of a trigger input, 16 opto-electric converters, 16 monitor signal outputs, the Digital Switch Control circuit (DSC), 16 daughter switch boards and two summing stages. The overall size is 370 mm (9 U) \* 220 mm \* 30 mm (6 HP).*

One multiplexer module consists of one 6 layer *motherboard* and 16 double layer *switchboards*, which are plugged into the *motherboard* via multiple pin connectors. Figure 5.6 shows a photo of the printed circuit MUX *motherboard* with 16 mounted daughter *switchboards*.

The *motherboard* includes the following components:

- 16 opto-electrical converters
- 16 monitor outputs
- the Digital Switch Control circuit (DSC)
- the trigger input to activate the DSC
- 16 ultra-fast MOSFET (metal-oxide-semiconductor field-effect transistor) switches on 16 *switchboards*
- the 16 channel summing stage.

One opto-electrical converter consists of a receptacle, a PIN photo diode, packed in the E2000-connector. The photodiode is biased by 12 V to reduce its intrinsic capacity for speed and noise optimization. The current signal of the PIN photo diode is converted into an equivalent voltage signal by a transimpedance-amplifier. Its amplifier-IC has a gain-bandwidth product of about 1.5 GHz and a very high slew rate of about 4000 V/$\mu$s.

## 5.2 The Ultra-fast Fiber-Optic MUX-FADC Data Acquisition System

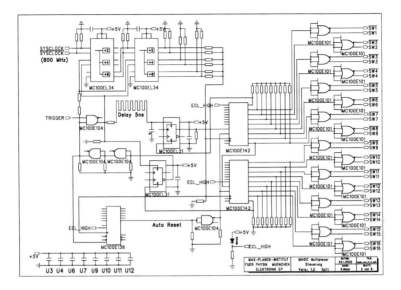

Figure 5.7: *Circuit diagram of the Digital Switch Control circuit (DSC): The trigger initiates a sequence of 16 PECL high levels of 40 ns duration applied consecutively to the switch boards.*

The trans-impedance is 1000 Ω. A monitor output consists of an ultra-wide band (UWB)-driver-amplifier, which transmits the signal from the transimpedance-stage to a 50 Ω SMA (SubMiniature version A) connector.

Figure 5.7 shows the circuit diagram of the Digital Switch Control circuit, DSC. It consists of the following parts:

- One clock generator-IC. It is programmable with a resolution of 12 bit from 50 MHz to 800 MHz and works in positive emitter coupled logic (PECL) mode. It is crystal stabilized and set to 800 MHz.

- A digital delay line (DDL) that can be set from 2 ns to 10 ns with 11 bit accuracy. It can be used to adjust the trigger times between different MUX *motherboards*.

- A digital lock-in-circuit (DLC) synchronizes the MUX-sequence to the trigger signal. The lock-in jitter is 1.25 ns (= 1/[800 MHz]).

- 16 differential PECL-drivers that transmit the MUX-sequence signals to the corresponding *switchboards*.

Each switchboard includes two ultra-wideband (UWB)-amplifier circuits, followed by two ns-switching MOSFETs operated in series and one UWB-driver-amplifier-circuit. MOSFET switches were chosen due to their fast switching properties and a very fast stabilization of the signal baseline after the switching. The small cross-talk through the closed switch is further reduced by the serial operation of two switches. An on-board PECL to CMOS converter distributes the digital switch-control-circuit (DSC)-signal to the MOSFET-switches in parallel. Figure 5.8 shows a photo of the switch board, while its circuit diagram is shown in figure 5.9.

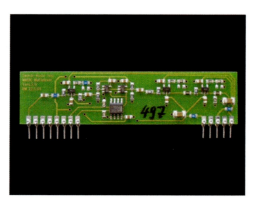

Figure 5.8: *The printed circuit board for fast switching. The switch board contains two ns-MOSFET-switches operated in series. Its mechanical dimensions are 80 mm * 20 mm *5 mm.*

In a passive summation, the switch parasitic capacitances would add up and can significantly widen the signal pulse. To avoid this, a two-step active summation was chosen: In the first step, the outputs of four channels are summed together. In the second step, the four resulting outputs are summed into one. For the summing UWB-amplifiers are used. The two-step setup keeps the channel wires short, and permits to use the amplifiers in the faster inverting mode while keeping the signal polarity non-inverting. Finally, an UWB-driver sends the multiplexed signals over a 50 $\Omega$-SMA-coaxial connection to the FADC. The circuit diagram of the summation stage is shown in figure 5.10.

### 5.2.4 FADC Read-Out

The FADCs are commercial products manufactured by the company Acqiris[2] (DC 282). They feature a 10 bit amplitude resolution, a bandwidth of 700 MHz, a sampling speed of 2 GSamples/s, an input voltage range of 1 V and an RMS noise level below 1.2 least significant bits (LSB). Each FADC board contains 4 channels. The read-out data are stored

---

[2]http://www.acqiris.com

## 5.3 Performance of the System Components

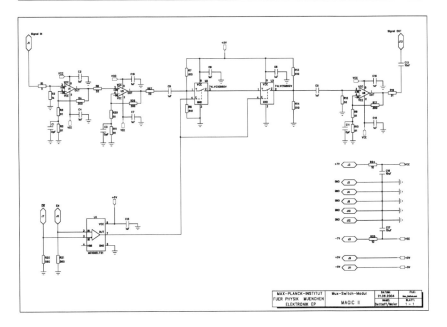

Figure 5.9: *Schematics of the fast switches: A high PECL level from the Digital Switch Control circuit opens both of the ns-MOSFET switches operated in series.*

in a random access memory (RAM) on the FADC board of 256 kSamples (512 kbytes) size per channel. Up to 4 FADC boards can be arranged in one compact Peripheral Component Interconnect (PCI) crate and are read out by a crate controller PC running under Linux. The FADCs are designed for a 66 MHz 64 bit data transfer via the compact PCI bus.

The FADC features a trigger time interpolator time to digital converter (TDC) that can be used to correct for a potential trigger jitter of 500 ps due to the asynchronous FADC clock with respect to the trigger decision. Table 5.2 summarizes the specifications of the ultra-fast FADCs.

## 5.3 Performance of the System Components

The performance of the MUX-FADC system components was studied in extensive laboratory tests: In section 5.3.1 the quality and performance of different commercially available optical splitters and delays is evaluated. In section 5.3.2 the performance of the multiplexing electronics is discussed. Finally, the quality and performance of the FADCs are tested in section 5.3.3.

# 5. The Data Acquisition System Upgrade of MAGIC

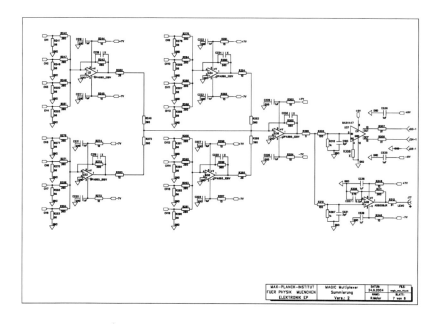

Figure 5.10: *Schematics of the two-step summing stage of the multiplexed signals. The two-step setup keeps the channel-wires short and allows to use the UWB-amplifiers in the faster inverting mode while keeping the signal polarity non-inverting.*

## 5.3.1 Performance of the Optical Delays and Splitters

The fiber-optic delay lines have channel-specific delay times of 0...15 times 40 ns plus 640 ns common base delay to allow ample timing for the trigger chain and internal system delays. Deviations from the specified delay times and potential changes in the delay due to temperature variations are important. It has to be ensured that all signals arrive in time at the multiplexer electronics when a given switch is open.

The manufacturer guarantees for the delay length a maximum deviation of ±2 ns from its nominal value. This was confirmed in laboratory measurements. Although the signal attenuation in fibers is small, nevertheless there are small differences in the dispersion of the signals in different channels due to the different delay lengths and connector tolerances.

Different technologies of fiber-optic splitters are available on the market. Three splitting technologies were tested: fused splitters, bifurcation splitters and so-called GRIN-splitters. In the fused technology two optical fibers are drilled and then thermally fused together. In bifurcation splitters the end faces of the two output fibers are mechanically attached

## 5.3 Performance of the System Components

| | |
|---|---|
| Mechanical size | 267 mm (6 U) * 220 mm * 30 mm (6 HP) |
| Number of channels | 4 |
| Analog input | 1 V full scale, adjustable offset |
| Sampling frequency | 2 GSamples/s |
| Sampling resolution | 10 bits |
| RAM size | 256 kSamples (512 kbytes) per channel |
| Bandwidth | 700 MHz |
| Noise level | < 1.2 LSB guaranteed |
| Power dissipation | 60 W (4 channels) |
| Trigger input | unipolar, adjustable threshold |

Table 5.2: *Specifications of the ultra-fast FADC.*

to the end face of the input fiber. In GRIN-type splitters the splitting is done by a semi-transparent mirror in conjunction with two graded index lenses (Lipson & Harvey 1983).

The MAGIC optical link uses multimode VCSELs and multimode optical fibers. Mechanical stress or deformations of the input fiber into the splitter, especially due to telescope movements, can vary the propagation of light modes in the fiber. The expected movements of the fibers were simulated in the laboratory by bending the fibers using different bending radii. The fused and bifurcation fiber-optic splitters show changes in the splitting ratio of more than $\pm 10\%$. Only the so-called GRIN type splitters are immune against mode changes, with measured changes of the splitting ratio of less than 1%.

The splitting ratio is guaranteed to be 50:50 within $\pm 3\%$ by the manufacturer. This was again confirmed in test measurements. All tested splitters were found to be insensitive with respect to time and temperature changes.

### 5.3.2 Performance of the MUX Electronics

The MUX electronics was extensively tested in the laboratory. For the use as a read-out system for MAGIC, the following points are very important:

- short switching noise and flat signal base-line
- high bandwidth (low pulse dispersion in amplitude and in time)
- strong signal attenuation for closed switches
- good linearity and large dynamic range
- stability.

Figure 5.11 shows a photo recorded with a fast oscilloscope of two consecutively multiplexed signals along with the switching noise between two channels. Although the switching noise is as large as 100 mV, it is very reproducible and confined to less than 10 ns of the 40 ns window per read-out channel. In the central part of the window the baseline is flat and stable.

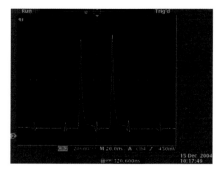

Figure 5.11: *Oscilloscope photo of two consecutive multiplexed signals and the baseline of two empty signal gates.*

The switching and summing stages only slightly widen the fast input pulses. A pulse of about 2.5 ns FWHM after the receiver photo diode is widened to 2.7 ns FWHM at the output of the multiplexer electronics. The pass-through of such fast signals through closed switches is less than 0.1% (60 dB attenuation).

Figure 5.12a shows the combined linearity of the switches and of the summing stage. The output signal amplitude of the MUX-board is plotted as a function of the input signal amplitude after the PIN-Diode, as measured at the monitor output. The right panel of the figure shows the deviations from linearity of the MUX-electronics. For output signals up to 1 V the MUX-electronics is linear with differential deviations less than 2%. The total non-linearity of the read-out chain is dominated by the analog optical link. Its response deviates from a perfect linear behavior by less than 10% in a total range of 56 dB (Paneque et al. 2003).

### 5.3.3 FADC Performance

The main performance parameters of the FADC are

- noise level / effective dynamic range
- linearity and bandwidth
- maximum trigger and acquisition rate
- dead time.

A noise level of less than 1.2 least significant bits (LSBs) is guaranteed by the manufacturer (700 MHz bandwidth, no input amplifier). There are small but constant differences in the input voltage full scales and thus in the gain for different FADC channels. These can be corrected for by the offline calibration software. The FADCs feature an internal calibration system keeping their integral and differential non-linearity below one LSB.

## 5.3 Performance of the System Components

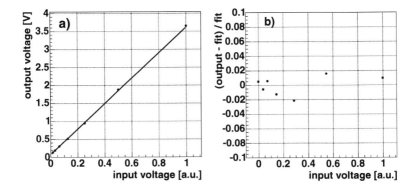

Figure 5.12: *a) Output signal amplitude of the MUX board as a function of the input signal after the PIN diode as measured at the monitor output. b) Residuals of the linearity of the MUX board as a function of the input signal after the PIN receiver diode as measured at the monitor output. The deviations are less than 2%.*

For the maximum trigger rate and the dead time optimization the interplay between the FADC boards and the crate controller PC is important. The compact PCI (cPCI) bus allows an effective data throughput of up to about 400 Mbytes/s (66 MHz, 64 bit) shared between all FADC channels in one crate.

In each event 2560 bytes are stored per FADC channel (16 channels of 40 ns gate time, 2 GSamples/s and 2 bytes per sample for the 10-bit resolution FADC). Reading out 8 FADC channels with one crate controller board results in a data volume of about 20 kbytes/event, which has to be transferred via the cPCI bus.

The ultra-fast FADC offers three modes of data acquisition:

- single acquisition
- segmented memory
- asynchronous acquisitions using a FIFO memory.

In the single acquisition mode the FADC writes the digitized data into the on-board RAM using it as one big ring buffer. Upon the arrival of a trigger the $N = 2560$ bytes corresponding to the event are copied to the PC with a direct memory access (DMA) transfer via the cPCI bus. The read-out time $T_1$ per event and per FADC channel in the crate is given by the sum of the DMA overhead time, $\text{Ovhd}_{\text{DMA}} \leq 25\,\mu\text{s}$ and the data transfer time over the cPCI bus:

$$T_1 = \text{Ovhd}_{\text{DMA}} + N \cdot \text{Xfr} \ . \tag{5.1}$$

$\text{Xfr} = 2.5\,\text{ns/byte}$ is the data throughput of the cPCI bus for the 64 bit, 66 MHz operation.

For 8 FADC channels per crate this amounts to a total transfer time of about 250 µs. Including a $\leq 25\,\mu$s global trigger rearm time this leads to about 275 µs dead time per event.

In the segmented mode the FADC RAM is divided into many segments. Each segment is used as a circular buffer where the digitized data is stored. After the arrival of a trigger the digitizer continues to write into the next segment. The dead time between two events is $\leq 25\,\mu$s. When all segments are filled the data is copied to the PC via one DMA transfer. The total read-out time for M segments is:

$$T_2 = \text{Ovhd}_{\text{DMA}} + M \cdot (N + \text{Extra}) \cdot \text{Xfr} , \qquad (5.2)$$

where Extra $\leq 200$ denotes the number of "overhead" data points per segment.

In the current scheme an additional time to reorder the data inside the PC has to be taken into account. For 8 FADC channels in one crate and 100 segments the total dead time amounts to $\leq 16$ ms, i.e. $\leq 160\,\mu$s per event.

The most attractive operation mode uses the FADC RAM as a FIFO (First In, First Out) memory. In this case the FADC writes the digitized data in one part of the RAM while previously stored data is asynchronously transferred to the PC. The dead time is thus reduced to the $\leq 25\,\mu$s needed to rearm the trigger. Additional dead time arises only if the average trigger rate exceeds the maximum sustainable rate of 2 kHz.

## 5.4 Prototype Test in the MAGIC Telescope on La Palma

Two prototype MUX-FADC read-out modules for 32 channels were tested as a read-out of the MAGIC telescope during two weeks in August/September 2004.

The main goals for the tests were:

- test of concept of the ultra-fast MUX-FADCs under realistic conditions

- study the interplay of the MUX-FADC system with the MAGIC trigger and data acquisition system

- implement the reconstruction and calibration for the ultra-fast digitized signals in the common MAGIC software framework MARS (Bretz & Wagner 2003)

- provide input for detailed MC simulations for the ultra-fast digitization.

This section is structured as follows: First, in section 5.4.1 the setup of the prototype test is presented. Thereafter, in section 5.4.2 an overview of the data taken is given. Finally, the data are analyzed and the test results are presented in section 5.4.3.

## 5.4 Prototype Test in the MAGIC Telescope on La Palma

### 5.4.1 Setup of the Prototype Test

Two MUX boards of 16 channels each were integrated into the MAGIC read-out system allowing the simultaneous data taking with the current 300 MSamples/s read-out and the MUX-FADC prototype read-out. Figure 5.13 shows in a block diagram how the MUX-FADC prototype read-out system was integrated into the current MAGIC FADC read-out. The analog optical signals arriving from the MAGIC PMT camera were split into two equal parts using fiber-optic splitters. One part of the optical signal was connected to the current MAGIC receiver boards which provided output signals to the MAGIC majority trigger logic, see section 2.2.5, and to the current 300 MSamples/s FADCs. The other part of the optical signal was delayed by a channel specific delay of 0...15 times 40 ns plus common base delay and directed to the optical receivers on the MUX boards. The common MAGIC trigger was used to trigger the MUX boards as well as the fast FADCs.

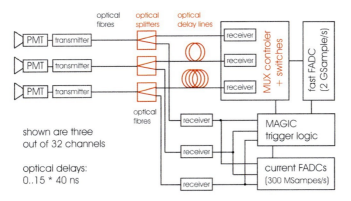

Figure 5.13: *Block diagram of the integration of the MUX-FADC prototype read-out in the current MAGIC FADC read-out. Shower images are simultaneously recorded with the ultra-fast MUX-FADC system and with the current MAGIC FADCs. The MAGIC trigger logic provides a trigger for the MUX electronics as well as for the ultra-fast FADCs.*

Figure 5.19 shows the group of 32 selected channels of the MAGIC PMT camera (Baixeras et al. 2004) to be read out by the ultra-fast digitizing system. The channels were chosen to be close packed in order to contain (at least partially) images of showers. In the test 16 bifurcation and 16 GRIN type splitters were used.

In order to acquire only events where the shower image is mostly located in the 32 MUX-FADC channels, only these channels were enabled in the MAGIC trigger system. The trigger fires if the signal in at least four close packed pixels exceeds the preset threshold.

In the prototype tests on La Palma an older version of the ultra-fast FADC was used, the Acqiris DC 240. It features a sampling speed of 2 GSamples/s with an 8 bit resolution. It was connected via a PCI bridge to a host PC running under Windows.

For every trigger 1300 FADC samples (16 times 80 samples plus 20 extra samples) were recorded with both of the used multiplexed FADC channels. An FADC memory of 120 segments was used. In the host PC the data were written into a binary file. This setup was chosen for simplicity and was not optimized for the smallest dead time in a continuous data taking mode. Nevertheless, the dead time between two of the 120 consecutively recorded events in the segmented mode was negligible.

### 5.4.2 The Data

The tests of the MUX-FADC system were carried out around the full moon period. In total about 230000 triggers were taken with the MUX-FADC read-out system (including pedestals and calibration LED light pulses). Table 5.3 summarizes the amount of data taken with and without the presence of moon light.

| trigger type | current FADC read-out | MUX-FADC read-out |
| --- | --- | --- |
| pedestals, no moon | 500 | 26400 |
| pedestals, moon | 5000 | 13210 |
| calibration, no moon | 47000 | 96000 |
| calibration, moon | 91000 | 70420 |
| cosmics, no moon | 500 | 8040 |
| cosmics, moon | 0 | 16800 |
| total | 144000 | 230870 |

Table 5.3: *Overview of the data taken during the MUX-FADC prototype test in the MAGIC telescope at La Palma.*

### 5.4.3 Data analysis

Each data file contains 120 events of two ultra-fast FADC channels with 1300 recorded FADC samples per event. The recorded raw data are converted into the usual ROOT-based MAGIC raw data format (Bretz & Wagner 2003), which provides the flexibility to adjust the number of recorded samples for each pixel.

#### 5.4.3.1 Signal Reconstruction

For each event the signals of 16 PMTs of the MAGIC camera are sequentially digitized by one FADC channel. As an example, figure 5.14 shows the raw data for 120 superimposed randomly triggered pedestal events. Between every two consecutive channels the switch noise is visible.

For calibration purposes the MAGIC PMT camera can be uniformly illuminated by a fast LED light pulser located in the center of the telescope dish, see section 2.2.7. Figure 5.15 shows the raw data of 16 consecutively read-out channels for 120 superimposed calibration events. The calibration signal pulses are clearly visible on the signal baseline. The gain difference from channel to channel is mainly due to a spread in the gain of the

## 5.4 Prototype Test in the MAGIC Telescope on La Palma

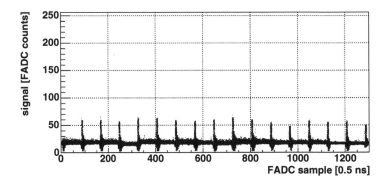

Figure 5.14: *Pedestals: Raw data (1300 samples for 16 consecutive channels) for 120 randomly triggered events superimposed.*

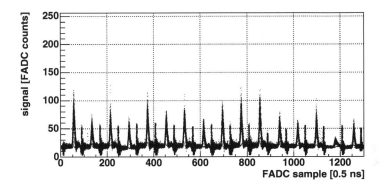

Figure 5.15: *Calibration events: Raw data (1300 samples for 16 consecutive channels) for 120 LED light pulses superimposed.*

VCSEL and receiver diodes of the analog optical link. The additional spread due to small differences in the fiber-optic splitters and a signal attenuation in the delay lines is small.

For each channel the pedestal level and pedestal RMS are calculated from either a pedestal run with random triggers or directly from the data. For the pedestal calculation a fixed number of FADC samples at a fixed position in the digitization window is used.

For the signal reconstruction a fixed number of FADC samples is integrated. The integration interval was chosen to be 4 FADC samples (corresponding to 4*3.33 ns =

# 5. The Data Acquisition System Upgrade of MAGIC

13.33 ns) for the current MAGIC FADCs. For the MUX-FADCs a window size of 10 FADC samples is chosen, corresponding to a 5 ns integration window. The reconstructed signal $\overline{S}$ is then given by:

$$\overline{S} = \sum_{i=i_0}^{i=i_0+3(9)} S_i \, , \qquad (5.3)$$

where $S_i$ is the i-th FADC sample after the trigger. The signal arrival time relative to the first FADC sample after the trigger, $t_{\text{arrival}}$, is reconstructed as the first moment of the FADC time samples used to calculate the reconstructed signal (see also section 3.1.4.2):

$$t_{\text{arrival}} = \frac{\sum_{i=i_0}^{i=i_0+3(9)} S_i(t_i - t_{i_0})}{\sum_{i=i_0}^{i=i_0+3(9)} S_i} \, . \qquad (5.4)$$

### 5.4.3.2 Calibration

The MAGIC camera can be homogeneously illuminated by fast LED light pulsers of different colors and intensities for calibration purposes, see section 2.2.7. The common MAGIC calibration algorithms (see section 3.2.1) and software were successfully applied to the ultra-fast digitization.

Figure 5.16 shows the distribution of the mean number of photoelectrons per pixel reconstructed with the current 300 MSamples/s FADC system and the MUX-FADC system. The MAGIC camera was illuminated with UV calibration pulses of about 2.5 ns FWHM. As expected, the mean reconstructed number of photo electrons is the same for the 32 split channels used in the MUX-FADC tests as for all the other MAGIC read-out channels.

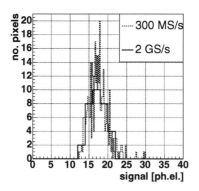

Figure 5.16: *Distributions of the mean reconstructed number of photoelectrons in the PMTs of the MAGIC camera from the LED pulser for the current 300 MSamples/s FADCs and the MUX-FADCs. Both read-out systems yield the same average number of photoelectrons.*

## 5.4 Prototype Test in the MAGIC Telescope on La Palma

Small differences in the cable length of the MAGIC analog optical link, the fiber-optic delays and transition times in the PMTs introduce arrival time differences between the pulses in different read-out channels of up to a few ns. These relative channel to channel time differences can also be calibrated using the LED pulser. One can determine the mean time difference between all pixels with respect to a reference pixel. In the calibration procedure of the cosmic events this timing difference is corrected for.

In addition, the event to event variation of the timing difference between two read-out channels for the LED pulser provides a measure of the timing accuracy. Figure 5.17 shows the distributions of the determined timing resolution of the current 300 MSamples/s FADCs together with the timing resolution of the MUX-FADCs. The timing accuracy strongly depends on the signal to noise ratio and the width of the input light pulse. The MUX-FADCs yield a better timing resolution by more than a factor of three compared to the current FADC system using the simple and stable timing extraction algorithm of equation (5.4).

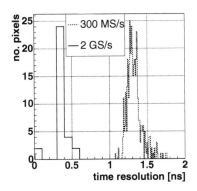

Figure 5.17: *Distributions of the timing resolution for the current 300 MSamples/s FADC read-out and the MUX-FADC read-out. The MUX-FADC system yields an improvement in the timing resolution by more than a factor of three.*

### 5.4.3.3 Cosmic Data

Cosmic shower data were recorded to study in detail the interplay of the ultra-fast MUX-FADC system with the MAGIC trigger logic. It also provides valuable input for the MAGIC MC simulations of the ultra-fast digitization system, e.g. about the pulse shapes for cosmic events.

In figure 5.18a one can see the pulse shape in a single pixel for a typical cosmic event. By overlaying the recorded FADC samples of many events after adjusting to the same arrival time, the average reconstructed pulse shapes can be calculated, see also section 3.1.2. Figure 5.18b shows the comparison of the average reconstructed pulse shapes recorded

180    5. The Data Acquisition System Upgrade of MAGIC

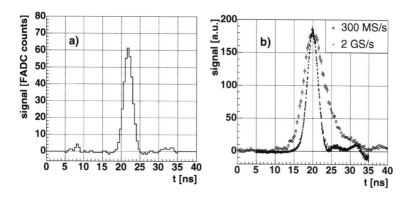

Figure 5.18: *a) Pulse shape in a single pixel for a typical cosmic event after pedestal subtraction. b) Comparison between the mean reconstructed pulse shapes recorded with the current MAGIC FADCs (open circles) and with the MUX-FADCs (full points).*

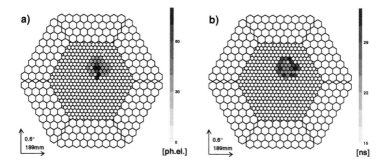

Figure 5.19: *a) Reconstructed signal in photoelectrons in the MAGIC PMT camera display and b) calibrated arrival times in ns in the MAGIC camera display for a typical cosmic event.*

with the current 300 MSamples/s MAGIC FADCs, including the 6 ns pulse stretching, and with the MUX-FADCs. The average reconstructed pulse shape for cosmic events has a FWHM of about 6.3 ns for the current FADC system and a FWHM of about 3.2 ns for the MUX-FADC system.

Figure 5.19a shows a MAGIC PMT camera display with the reconstructed signal af-

## 5.4 Prototype Test in the MAGIC Telescope on La Palma

ter calibration in photoelectrons for a typical cosmic event. For the same event figure 5.19b shows the reconstructed arrival time after correction for the channel-to-channel time differences.

#### 5.4.3.4 Pedestals / Noise

In the data recorded by an IACT, the electronics noise together with the LONS fluctuations is superimposed on the Cherenkov signal from air-showers. The noise from the LONS can be simulated as the superposition of the detector response to single photoelectrons, arriving at a given rate but randomly distributed in time. This can be quantified using the noise autocorrelation function $B_{ij}$, the correlation between the read-out samples $i$ and $j$ (see also section 3.1.4.4):

$$B_{ij} = \langle b_i b_j \rangle - \langle b_i \rangle \langle b_j \rangle , \qquad (5.5)$$

where $b_i$ and $b_j$ are the FADC samples $i$ and $j$ for a pedestal event.

Figure 5.20 shows the noise autocorrelation for the current MAGIC FADC system and the MUX-FADC system with open camera (i.e. exposed to LONS), normalized to the pedestal RMS. In the same plot, the noise autocorrelation for the MUX-FADC system with closed camera (no LONS), normalized to the pedestal RMS for an open camera, is shown. The noise autocorrelation of the current FADC system extends to several ns since the pulse is stretched by 6 ns. For the MUX-FADC system with no pulse shaping there is still a considerable noise autocorrelation for an open camera. The noise autocorrelation mostly disappears in case of a closed camera with electronics noise only.

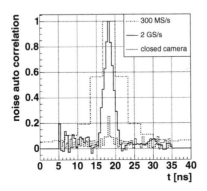

Figure 5.20: *Noise autocorrelation function with respect to a fixed FADC sample for the current MAGIC read-out chain with 6 ns pulse shaping, the MUX-FADC read-out with open camera and closed camera, normalized to the pedestal RMS of the opened camera.*

Figure 5.21 shows the distributions of the integrated noise (integrated pedestal RMS after calibration in photoelectrons) per pixel for the current FADC system and for the

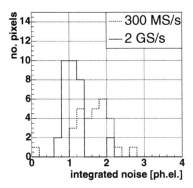

Figure 5.21: *Distributions of the integrated noise per pixel in the signal reconstruction window after calibration into photoelectrons for the current 300 MSamples/s FADC read-out and using the MUX-FADC read-out.*

MUX-FADC system. The shorter integration time used for the pulse reconstruction with the MUX-FADC system yields a reduction of the effective integrated noise by about 40%.

Using the new MUX-FADC system the noise contributions due to the LONS may even be resolved into individual pulses. Figure 5.22 shows a typical example for the signals in a pedestal event (random triggers). The pedestal does not vary in an uncorrelated way. Instead most of the pedestal variations are due to peaks on the signal baseline.

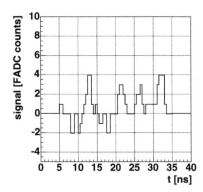

Figure 5.22: *Time structure in a typical pedestal event. The peaks on the baseline are most likely due to single photo electrons from the light of the night sky.*

## 5.4 Prototype Test in the MAGIC Telescope on La Palma

The rate of the peaks was studied to verify whether it is compatible with the rate of LONS photoelectrons. A window of 6 slices is slid over the FADC samples of randomly triggered pedestal events. The first window position after the switch noise where the sum of the FADC samples exceeds the pedestal level by at least 3 FADC counts was chosen. Figure 5.23a shows the arrival time distribution of the first noise peak. The distribution can be fitted by an exponential function with a rate $r$ of

$$r = (0.13 \pm 0.01) \text{ns}^{-1} \ . \tag{5.6}$$

This corresponds to an integrated LONS charge of about 1.3 photoelectrons per 10 ns integration window, which is in good agreement with the expected LONS rate.

Figure 5.23b shows the pulse shape of the selected noise peak averaged over many events. It has a FWHM of about 2.6 ns. This corresponds to the response of the instrument to a $\delta$-function input pulse (single LONS photoelectrons have no internal time structure). The mean charge of the noise peak corresponds within errors to the mean charge for a single photoelectron.

The 8 bit amplitude resolution in the test setup was somewhat limiting the resolution of the single photoelectrons due to the LONS. With the higher resolution of 10 bit with the full MUX-FADC system even a continuous calibration of the read-out chains using the single photoelectrons shall be possible.

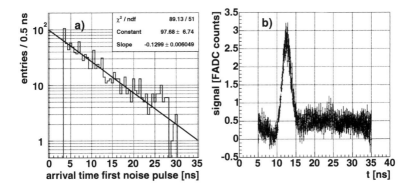

Figure 5.23: *a) Arrival time distribution of the first noise peak on the pedestal baseline. The peaks are arriving randomly in time with a rate of $(0.13 \pm 0.01)\text{ns}^{-1}$. b) Average reconstructed shape of the LONS noise peaks. The FWHM is about 2.6 ns.*

### 5.4.3.5 MC Simulations

The response of the MAGIC telescope to $\gamma$-ray showers and to background was simulated in detail, see section 2.2.8 and Majumdar et al. (2005). Both the currently used 300 MSamples/s read-out chain and the ultra-fast digitization were simulated.

Figure 5.24 shows the reconstructed single photo electron spectrum of a simulated pedestal run. The highest integral of 8 FADC slices (4 ns) was searched for in a fixed digitization window of 20 slices (10 ns). The leftmost peak corresponds to electronics noise only. The right part of the distribution corresponds to the response of the PMT to one or more photo electrons.

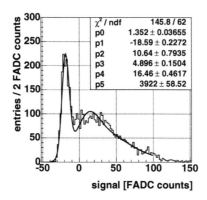

Figure 5.24: *Reconstructed single photoelectron spectrum of a simulated pedestal run. The leftmost peak is the pedestal.*

In figures 5.25 the signal and arrival time resolutions of the current and the MUX-FADC system are compared using MC simulations. For both MC simulations the same LONS conditions are assumed as well as the same electronics noise level. In the simulation the intrinsic transit time spread of the PMTs of about 250 ps per photoelectron was not taken into account. The input light pulse has a FWHM of 1 ns as expected for $\gamma$-ray induced showers.

Contrary to the simple signal and arrival time extraction algorithms used above, a dedicated numerical fit to the FADC samples using a known pulse shape has been applied to illustrate the theoretically achievable resolution, see section 3.1.4.4. Figure 5.25a shows the resolution of the reconstructed pulse arrival time as a function of the input signal. The MUX-FADC system improves the timing resolution by nearly a factor of two. Figure 5.25b shows the resolution of the reconstructed charge as a function of the input charge. With the MUX-FADC system the charge resolution improves by a factor of two. It should be noted that in MAGIC II an even higher resolution could be achieved because in the MAGIC I camera the PMT signals were already stretched at the VCSEL driver amplifiers in order to prevent an oscillation. In MAGIC II a better amplifier should allow to transmit the full bandwidth of the PMTs.

## 5.5 Discussion

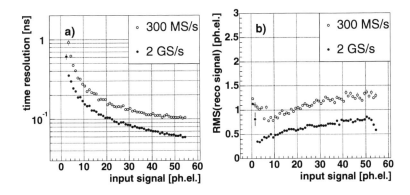

Figure 5.25: *MC simulations: a) Comparison of the pulse arrival time resolution as a function of the input signal size between the current MAGIC 300 MSamples/s FADCs and the 2 GSamples/s FADCs. The time resolution improves by nearly a factor of two with the new system. b) Comparison of the signal resolution as a function of the input signal with the current 300 MSamples/s MAGIC FADCs and the 2 GSamples/s FADCs system. The signal resolution improves by a factor of about two.*

## 5.5 Discussion

The newly developed ultra-fast fiber-optic multiplexed FADC prototype read-out system was successfully tested during normal observations of the MAGIC telescope in La Palma. The fiber-optic splitters and delays are commercially available and comply with the required specifications for the use in the ultra-fast MUX-FADC read-out system. The 10 bit 2 GSamples/s FADCs from Acqiris have been developed for MAGIC, and are available now as a commercial product. Thus the ultra-fast FADC read-out has grown to a viable technology which is ready for the use as a standard read-out system for the MAGIC telescope and other high-speed data acquisition applications.

The multiplexing of 16 channels into one ultra-fast FADC allows one to reduce the price of an ultra-fast read-out system. The MUX-FADC read-out reduces the costs by about 85% compared to using one ultra-fast FADC per read-out channel. Also the power consumption of the read-out system is reduced by at least a factor of 10.

The ultra-fast MUX-FADC system allows to skip the artificial pulse stretching and thus to use a shorter integration window for the Cherenkov pulses. The reduction of the pulse integration window from 13.33 ns (4 samples with 3.33 ns per sample) for the current MAGIC FADC system to 5 ns (10 samples with 0.5 ns per sample) for the MUX-FADC system corresponds to a reduction of the integrated LONS charge by a factor of about 2.7. Consequently, the RMS noise of the LONS is reduced by about 40%.

The recorded images of the air showers are usually, at least for energies above 100 GeV, characterized by so-called Hillas parameters, see section 3.2.4. In order to prepare raw shower images for the Hillas parameter calculation it is necessary to apply so-called tail cuts to reject pixels with a low signal-to-noise ratio, see section 3.2.3. All pixels with signals below 3 times their noise RMS (mainly due to LONS) are rejected (dynamic image tail cut cleaning).

A reduction in the noise RMS translates into lower image cleaning levels. Thus a larger part of the shower image, or in other words a shower image of a higher signal-to-noise ratio, can be used to calculate Hillas parameters. This is especially important for low energy events where the signals of only a few pixels are above the image cleaning levels. The image quality of the air showers will improve with the ultra-fast read-out system. This will allow the reduction of the analysis energy threshold of the MAGIC telescope.

The ultra-fast FADC system also provides an improved resolution of the timing structure of the shower images. As indicated by MC simulations (Mirzoyan et al. 2006) $\gamma$-ray showers, cosmic ray showers and the so-called single muon events have different timing structures. Thus the ultra-fast FADC read-out should enhance the separation power of $\gamma$-ray showers from backgrounds.

## 5.6 Production and Installation of the Full MUX-FADC System

After the successful prototype tests of the MUX-FADC system the Max-Planck society has allocated special funds to build and install the MUX-FADC system as a read-out upgrade of the full MAGIC telescope (end of 2004). In 2005 the MUX electronics was produced and tested in the MPI, the optical components were purchased and their quality controlled as well as the Acqiris FADCs. Figure 5.26 shows the full read-out system consisting of four racks in the MPI electronics department during tests.

In spring 2006 the full MUX-FADC system was installed in the MAGIC telescope and data are being recorded since May 2006 in parallel to the 300 MSamples/s FADC read-out. The analysis of these data is beyond the scope of this thesis. Figure 5.27 shows the installed MUX-FADC racks in the electronics room of the MAGIC counting house in La Palma.

## 5.6 Production and Installation of the Full MUX-FADC System

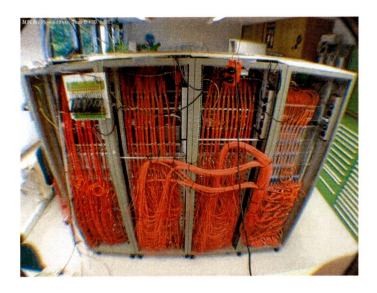

Figure 5.26: *Full MUX-FADC read-out system assembled in the electronics department at MPI during quality control tests. The racks are shown from the back side with preliminary connected optical fibers.*

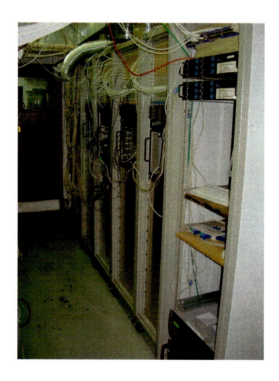

Figure 5.27: *Full MUX-FADC system after installation in the electronics room of the MAGIC counting house in La Palma.*

# Chapter 6
# Summary, Conclusion and Outlook

This thesis had the following four main aims:

1. Observation of very high energy $\gamma$-rays from galactic sources with the MAGIC telescope. Modelling and discussion of the multi-wavelength emission of the sources.

2. Development of dedicated observation and analysis procedures for large zenith angles. Implementation of the off-source tracking observation mode (the so-called wobble observation mode). Computation of VHE $\gamma$-ray sky maps and comparison with multiwavelength data.

3. Development and implementation of an FADC pulse reconstruction algorithm in the common MAGIC analysis software framework MARS.

4. Development of new ultra-fast read-out electronics for the upgrade of the MAGIC telescope, including prototype-tests, production and installation of the full-scale system.

In the first half of my time as a PhD student, I have developed, in cooperation with the electronics department of the MPI, an "Ultra Fast Fiber-Optic Multiplexed FADC system." In summer 2004 I performed a successful test of a prototype of this new read-out system at the MAGIC telescope. The results are published in Nuclear Instruments and Methods A (Bartko et al. 2005b). Based on this prototype test the Max-Planck-Gesellschaft has granted the MAGIC group funds to upgrade the current MAGIC read-out system with the newly developed one. I have supervised the production and quality control of the hardware of the new read-out system till fall 2005. Together with colleagues from the MPI we installed the full new read-out system in the MAGIC telescope at La Palma in April 2006, followed by a test and commissioning phase. Moreover, we have adapted the MAGIC read-out and data analysis software to the new data format.

Since May 2006 data are being recorded with the new MUX-FADC system in parallel to the 300 MSamples/s FADC read-out. The analysis of these data is beyond the scope of this thesis. It will be finalized after submission of this thesis. The parallel read-out with the old and the new data acquisition system will allow a reliable intercalibration of the two systems. In the near future, the MUX-FADC system itself will have to be updated in the process of the installation of the second MAGIC telescope, see section 2.4.

# 6. Summary, Conclusion and Outlook

On the software side, I have developed a novel FADC pulse reconstruction algorithm for Cherenkov telescopes. It computes the signal charge and the arrival time from the recorded FADC samples in each camera channel, for each triggered Cherenkov light pulse. Based on the features of the current read-out electronics, the reconstruction algorithm performs a numeric fit of a model signal shape to the recorded signal samples varying the amplitude and the position in time (the "digital filter" method). The full noise autocorrelation is taken into account. I have presented the performance of the algorithm at an international conference (Bartko et al. 2005a), and a publication is submitted to a refereed journal (Albert et al. 2006e). The reconstruction algorithm has been implemented in the common MAGIC software framework.

The digital filter method will have to be adjusted to the data taken with the ultrafast MUX-FADCs. Because no pulse stretching is applied, differences in the pulse shape between $\gamma$-ray and hadron initiated showers are expected. The quality of the numerical fit of the expected $\gamma$-ray pulse shape to the recorded FADC samples may give, on an event by event basis, an additional parameter to discriminate $\gamma$-ray from hadron events. Furthermore, the improved arrival time information for the signals in each of the pixels may be further exploited to reject noise in the image cleaning (see section 3.2.3) as well as to improve the $\gamma$/hadron separation (see section 5.1).

After the completion of the commissioning phase of the MAGIC telescope, in the beginning of 2005 the cycle one observations started. For this first scientific observation cycle I have proposed two observations as Principle Investigator: the Galactic Center and some of the galactic sources recently detected in the galactic scan performed by the HESS collaboration (Aharonian et al. 2005a). The Galactic Center is a very interesting region containing most likely a super massive black hole, supernova remnants, pulsar wind nebula candidates, hot gas and large magnetic fields. Recently, very high energy $\gamma$-radiation was reported from the Galactic Center by the CANGAROO (Tsuchiya et al. 2004), VERITAS (Kosack et al. 2004) and HESS (Aharonian et al. 2004b) collaborations. The fact that the measured spectra differed significantly has stimulated discussions about the origin of the differences. The source of the very high energy $\gamma$ radiation is still unclear.

Most of the new galactic sources discovered by the HESS collaboration are part of a new population of galactic very high energy $\gamma$-ray sources. Though most of these sources are spatially coincident with supernova remnants (either from the shell type or pulsar wind nebulae), none of these new sources had been predicted to be observable in the very high energy $\gamma$-ray domain. In contrast to that, only few of the supernova remnants, which were predicted to emit very high energy $\gamma$-radiation, could actually be observed. For the theoretical understanding and modeling of these sources the energy spectrum, flux variations in time and the exact source location and extension were measured.

The Galactic Center and the new galactic HESS sources are located in the southern part of the sky. With the MAGIC telescope they can only be observed under large zenith angles (ZA) of up to 60°. These observations were the first observations with the MAGIC telescope at such large ZA. The shower images for observations at large ZA (see section 3.6.5) differ significantly from the ones for standard observations at small ZA. Therefore, I have developed and implemented the data analysis chain for large ZA data and tested it with data samples of the Crab Nebula as a bright and steady reference source. As the galactic star field results in quite an inhomogeneous background brightness distribution, I have decided to observe the sources in the so-called wobble observation mode (see section 2.3). These were the first scientific observations with the MAGIC telescope in the wobble observation mode. To analyze the wobble mode data I have updated the MAGIC standard analysis and reconstruction software (see section 3.5). These observations were also the first observations with the MAGIC telescope of potentially extended objects with uncertain sky positions. Therefore, I have developed a procedure to calculate an unbiased sky map of VHE $\gamma$-rays (see section 3.3.2) and to overlay this sky map with contours from observations in other wavelength bands.

In an extensive observation campaign in summer 2005 we have accumulated a total exposure time of about 50 hours on the Galactic Center, about 25 hours on the source HESS J1813-178 and another 20 hours on the source HESS J1834-078. Some of this data taking I have supervised as shift leader of the MAGIC telescope. These observations resulted in significant detections of the three sources. Working together with theorists I have discussed the results considering models for the multi-wavelength emission of the sources. The results of the observations are published in three Letters to the Astrophysical Journal (Albert et al. 2006a,b,c). In the following sections 6.1 to 6.3 the main results of these observations are summarized and for each source an outlook is given, as to which further observations are needed for a better understanding of these sources. This concerns especially the parent particles of the VHE $\gamma$-rays and the acceleration models. Finally, section 6.4 discusses which measurements are necessary in the future to identify the sources of the cosmic rays and which implications these requirements have on future instrumentation.

## 6.1 The Galactic Center

The Galactic Center was observed with the MAGIC telescope under large zenith angles resulting in the detection of a differential $\gamma$-ray flux consistent with a steady, hard-slope power law, described as $\mathrm{d}N_\gamma/(\mathrm{d}A\mathrm{d}t\mathrm{d}E) = (2.9\pm0.6)\times10^{-12}(E/\mathrm{TeV})^{-2.2\pm0.2}$ cm$^{-2}$s$^{-1}$TeV$^{-1}$. This measurement confirms the VHE $\gamma$-ray source at the Galactic Center. Within errors the determined flux is in very good agreement with the measurement of HESS (Aharonian et al. 2004b). The VHE $\gamma$-ray emission does not show any significant time variability; the measurements rather suggest a steady emission of $\gamma$-rays from the GC region. The CANGAROO collaboration has revised their error estimation of the Galactic Center data such that their results are marginally compatible with the MAGIC and HESS VHE $\gamma$-rays spectra within errors (Katagiri et al. 2005). As the VHE $\gamma$-ray spectrum extends up to 20 TeV following a simple power law, the main part of the observed VHE $\gamma$-radiation is most probably not due to Dark Matter particle annihilation in the GC region, which would in most models result in a cut-off of the $\gamma$-ray spectrum well below 10 TeV.

There are still various candidate sources for the VHE $\gamma$-ray emission in the immediate vicinity of the Galactic Center: The MAGIC telescope data show a point like source, centered at (RA, Dec)=($17^\mathrm{h}45^\mathrm{m}20^\mathrm{s}$, -29°2′). Within the systematic pointing uncertainty of 2′ the source location is spatially consistent with the three candidates SgrA*, G359.95-0.04 as well as SgrA East.

In order to be able to discriminate between the different source models future experiments should provide the following features:

- Improved angular resolution: The projected distance between the objects SgrA*, G359.95-0.04 and SgrA East is below 1'. G359.95-0.04 and SgrA East have some intrinsic extension. Therefore, also a VHE $\gamma$-ray signal from these source is expected to be slightly extended, whereas the VHE $\gamma$-ray signal from the immediate vicinity of the black hole might be point-like to a very good approximation.

- Larger $\gamma$-ray spectral coverage: The expected $\gamma$-ray spectra from hadronic interactions and IC scattering of leptons are different, especially at energies around 1 GeV. An open question is, whether the VHE $\gamma$-ray source is physically connected to the EGRET source (Mayer-Hasselwander et al. 1998; Hartman et al. 1999) near the Galactic Center, observed in the energy region from 100 MeV to about 10 GeV. In case of cosmic ray interactions with dense target material the VHE $\gamma$-ray spectrum is expected to continue up to at least 100 TeV.

- Full time coverage of the VHE $\gamma$-ray source in order to study the possible connection of the VHE $\gamma$-ray emission with the flaring emission in the near infrared and X-ray bands.

- Further study of the connection between the diffuse $\gamma$-radiation along the Galactic Plane and the point-like VHE $\gamma$-ray source at the GC.

- Search for VHE neutrinos or ultra high energy ($\geq 10^{18}$ eV) neutrons from the GC to test the hadronic models for the production of the VHE $\gamma$-rays. Such high energy neutrons have a good chance to reach the earth before decaying.

## 6.2 HESS J1813-178

The newly discovered source HESS J1813-178, which is spatially coincident with the SNR G12.82-0.02 is observed with the MAGIC telescope resulting in the detection of a differential $\gamma$-ray flux consistent with a hard-slope power law, described as $dN_\gamma/(dAdtdE) = (3.3\pm0.5)\times10^{-12}(E/\text{TeV})^{-2.1\pm0.2}$ cm$^{-2}$s$^{-1}$TeV$^{-1}$. The observation is put in the perspective of multifrequency observations.

The source HESS J1813-178 is a very good candidate for an accelerator of hadronic cosmic rays. The current data can be described by both hadronic as well as leptonic models for the $\gamma$-ray emission. To distinguish between these two possibilities, data at around 1 GeV are needed, see figure 4.11. These data may soon be available from the GLAST satellite (see e.g. Wood et al. (1995)), expected to be launched in late 2007. The detection of TeV neutrinos from this source with future neutrino telescopes like IceCube (Ahrens et al. 2004) and KM3NeT (Katz 2006) would strongly support the hadron acceleration in this source. In addition to prove that the cosmic rays are accelerated in this source, one also has to show that hadrons are accelerated at least up to the knee energy of the cosmic ray spectrum of a few times $10^{15}$ eV. In this case $\gamma$-rays of energies of at least 100 TeV would have to be detected from this source. Possible instruments for this are the high energy Cherenkov Telescope Array CTA (Colin et al. 2006) or the future HAWK detector (Smith

2005). At such energies MAGIC measurements would be limited by $\gamma$-ray event statistics due to the small flux.

VHE $\gamma$-ray data with a higher angular resolution (for example in combined MAGIC I and MAGIC II stereo observations, see section 2.4) may be able to accurately determine the intrinsic source size and distinguish whether the $\gamma$-rays are produced at the SNR shell or in the interior of the SNR as expected in the case of a pulsar wind nebula. As shown in section 4.2.4.2, it is very difficult to attribute the radio emission, soft X-rays as measured by ASCA (Brogan et al. 2005) and hard X-rays as measured by INTEGRAL (Ubertini et al. 2005) as synchrotron emission from just one electron population. Future X-ray data with a high angular resolution will tell whether the X-ray emission originates from the shell of the remnant and/or from a synchrotron nebula in its interior. An intercalibration and independent confirmation of the ASCA and INTEGRAL source spectra is also highly desirable.

## 6.3 HESS J1834-087

The newly discovered source HESS J1834-087 is spatially coincident with the SNR G23.3-0.3 (W41). It is observed with the MAGIC telescope, resulting in the detection of a differential $\gamma$-ray flux consistent with a power law, described as $dN_\gamma/(dAdtdE) = (3.7 \pm 0.6) \times 10^{-12} (E/\text{TeV})^{-2.5\pm0.2} \text{cm}^{-2}\text{s}^{-1}\text{TeV}^{-1}$. A source extension of $(0.14\pm0.04)°$ is derived. A spatial coincidence of the $\gamma$-ray source with a massive molecular cloud observed by its $^{13}$CO and $^{12}$CO emission is found.

The $\gamma$-ray energy spectrum of HESS J1834-087 is steeper than the one expected from the prototype source of cosmic rays. Also the VHE $\gamma$-ray emission is centered in projection inside the SNR shell. Therefore, the observed VHE $\gamma$-rays may either be due to IC scattering of VHE electrons inside the SNR shell (a possible pulsar wind nebula) or due to the energy dependent diffusion of the accelerated hadronic cosmic rays to the densest region of the molecular cloud. Future high angular resolution observations of VHE $\gamma$-rays (for example with MAGIC II, see section 2.4) may be able to definitely localize the $\gamma$-ray production place. Sensitive X-ray data are needed to study whether there is a population of VHE electrons inside the SNR, which may be responsible for the observed $\gamma$-ray emission. Also the radio hot spot in the middle of the SNR shell should be investigated in more detail in radio and X-rays. More sensitive radio observations may detect or exclude the existence of a pulsar within the SNR shell.

## 6.4 Future Searches for the Accelerators of the Cosmic Rays

This thesis has shown that the sources HESS J 1813-178 and HESS J 1834-087 are good candidates for accelerators of the cosmic rays. The region of a few pc radius around the Galactic Center also probably hosts an accelerator of cosmic rays. Despite this progress it is still an open question, which classes of astrophysical objects are responsible for the acceleration of the cosmic rays. In particular, the following questions arise:

# 6. Summary, Conclusion and Outlook

- Are all accelerators of cosmic rays expected to emit VHE $\gamma$-rays and neutrinos, and if so, at what flux levels? Are some additional conditions required like the interaction with molecular clouds to provide a target for the $\gamma$-ray and neutrino production?

- Which classes of galactic objects emit VHE $\gamma$-rays?

- How can one determine for a particular source the relative fractions of VHE $\gamma$-rays, which are produced in hadronic interactions and which are produced in reactions of VHE leptons? The physical processes of particle acceleration and $\gamma$-ray emission in this source have to be determined for each of these sources separately.

- Do the candidate sources of the cosmic rays accelerate hadrons up to the energy of the knee in the cosmic rays spectrum? Which physical parameters of the source determine the knee energy?

- Are there enough candidate sources to account for the total observed power of the cosmic rays?

The following observations can contribute to answer these open questions:

- A high sensitivity unbiased scan of the galactic plane in VHE $\gamma$-rays would provide an overview and statistics of the types and properties of the candidate sources for cosmic ray acceleration. These $\gamma$-ray measurements require new generation IACTs with a higher flux sensitivity, smaller $\gamma$-ray PSF, larger field of view and higher duty cycle, like MAGIC II and CTA.

- In order to show that hadrons are accelerated up to the energy of the knee in the cosmic ray energy spectrum at about a few times $10^{15}$ eV in the source, the measurement of the $\gamma$-ray spectrum has to be extended up to at least 100 TeV. For $\gamma$-ray energies around 100 TeV large arrays of IACTs like CTA or future large water Cherenkov telescopes like HAWK are suited best.

- Good angular resolution data at around 1 GeV have to be collected on the candidate source to be able to distinguish the case of production of the observed VHE $\gamma$-rays in the collision of cosmic ray hadrons with ambient matter from the case of leptonic production. This observation can be carried out by future space-borne $\gamma$-ray telescopes like the future GLAST satellite.

- Sensitive soft and hard X-ray data have to be taken on the candidate sources in order to evaluate whether there are populations of VHE electrons which may produce the observed VHE $\gamma$-ray fluxes. Generally, cut-offs of the synchrotron spectrum are expected at a maximum of a few keV as electrons are cooling quite efficiently in the ambient galactic magnetic fields and may not be accelerated beyond a few tens of TeV. A very hard X-ray component may come from the synchrotron radiation of secondary electrons produced in hadronic reactions of the accelerated cosmic rays in or near the source. This physics objective requires sensitive soft and hard X-ray satellite telescopes with good spectral and spatial resolution.

## 6.4 Future Searches for the Accelerators of the Cosmic Rays

- The detection of TeV neutrinos from this source would strongly support the hadron acceleration in this source. Unfortunately, the neutrino interaction cross-section is minute, which requires km$^3$ detectors to observe a significant neutrino signal from galactic $\gamma$-ray sources, see e.g. Kappes et al. (2006). The central part of the galaxy can only be observed from neutrino telescopes in the northern hemisphere like KM3NeT (Katz 2006).

- Finally, one has to observe enough sources, in which cosmic rays are accelerated, such that the sum of the sources are capable to provide the full power observed in cosmic rays.

# List of Acronyms and Abbreviations

| | |
|---|---|
| ACT | Air Cherenkov Telescopes |
| ADAF | Advection Dominated Accretion Flow |
| AGN | Active Galactic Nucleus |
| AIROBICC | AIr-shower Observation By Angle Integrating Cerenkov Counters |
| asl | above sea level |
| b | galactic lattitude |
| CANGAROO | Collaboration of Australia and Nippon for a GAmma Ray Observatory in the Outback |
| CGRO | Compton Gammay Ray Observatory |
| CMB | Cosmic Microwave Background |
| COMPTEL | Compton Telescope |
| CORSIKA | COsmic Ray SImulations for KAscade |
| DAQ | Data Acquisition |
| DDL | Digital Delay Line |
| Dec | Declination |
| DLC | Digital Lock-in-Circuit |
| DM | Dark Matter |
| DMA | Direct Memory Access |
| DSC | Digital Switch Control |
| EBL | Extra-galactic Background Light |
| EGRET | Energetic Gamma Ray Experiment Telescope |
| eV | electron Volt |
| FADC | Flash Analog-to-Digital Converter |
| FIFO | First In, First Out |
| FIR | Far-InfraRed |
| FOV | Field-Of-View |
| FWHM | Full Width at Half Maximum |
| GC | Galactic Center |
| GLAST | Gamma-ray Large Area Space Telescope |
| GMC | Giant Molecular Cloud |
| GRB | $\gamma$-Ray Bursts |
| GRIN | GRaded INdex |

| | |
|---|---|
| HEGRA | High Energy Gamma Ray Astronomy |
| HESS | High Energy Stereoscopic System |
| HP | Horizontal Pitch |
| HPD | Hybrid Photo Detectors |
| HV | High-Voltage |
| IACT | Imaging Air Cherenkov Telescope |
| IC | Inverse Compton |
| INTEGRAL | INTErnational Gamma-Ray Astrophysics Laboratory |
| ISM | InterStellar Medium |
| l | galactic longitude |
| LED | Light-Emitting Diode |
| LONS | Light Of the Night Sky |
| los | line of sight |
| LSB | Least Significant Bit |
| $M_\odot$ | solar Mass |
| MAGIC | Major Atmospheric Gamma-ray Imaging Cherenkov |
| MARS | MAGIC Analysis and Reconstruction Software |
| MC | Monte Carlo (simulation) |
| MEGA | Medium Energy Gamma-Ray Astronomy |
| MJD | Modified Julian Date |
| MOSFET | Metal-Oxide-Semiconductor Field-Effect Transistor |
| mSUGRA | minimal Supergravity |
| MUX | multiplexed |
| n.d.f | number of degrees of freedom |
| NFW | Navarro, Frenk, & White |
| OSSE | Oriented Scintillation Spectrometer Experiment |
| pc | parsec, parallax of one arc second |
| PCI | Peripheral Component Interconnect |
| PECL | Positive Emitter Coupled Logic |
| PIN | P-type, Intrinsic, N-type |
| PMT | Photo Multiplier Tube |
| PSF | Point Spread Function |
| PWN | Pulsar Wind Nebula, plerion |
| QE | Quantum Efficiency |
| QED | Quantum Electro Dynamics |
| QCD | Quantum Chromo Dynamics |
| RA | Right Ascension |
| RAM | Random Access Memory |
| RF | Random Forest |
| RMS | Root Mean Square |
| ROSAT | Röntgen Satellite |

# Abbreviations

| | |
|---|---|
| Sgr | Sagittarius |
| SMA | SubMiniature version A |
| SN | SuperNova |
| SNR | SuperNova Remnant |
| SSC | Synchrotron-Self-Compton |
| SUSY | Super Symmetry |
| TDC | Time to Digital Converter |
| TT | Transit Time |
| TTS | Transit Time Spread |
| UV | ultra violet |
| UWB | Ultra-Wide Band |
| VCSEL | Vertical Cavity Surface Emitting Laser |
| VERITAS | Very Energetic Radiation Imaging Telescope Array System |
| VHE | Very High Energy |
| VLA | Very Large Array |
| WIMP | Weakly Interacting Massive Particle |
| ZA | Zenith Angle |

# List of Figures

| | | |
|---|---|---|
| 1.1 | Known VHE $\gamma$-ray sources in the sky. | 10 |
| 1.2 | Instrument sensitivities for $\gamma$-ray detection. | 11 |
| 1.3 | Cosmic ray spectrum. | 20 |
| 1.4 | Fermi acceleration. | 23 |
| 1.5 | Structure of the Milky Way. | 24 |
| 1.6 | Binary Systems. | 29 |
| 1.7 | Sources near the Galactic Center. | 32 |
| 1.8 | Sources near the Galactic Center, zoomed in. | 33 |
| 1.9 | $\gamma$-ray flux from GC. | 34 |
| 1.10 | VHE $\gamma$-ray source location at the GC. | 34 |
| 1.11 | Total spectrum of the GC. | 35 |
| 1.12 | Multiwavelength observations of the Milky Way. | 36 |
| 1.13 | Significance map for VHE $\gamma$-radiation from the inner galaxy. | 37 |
| 1.14 | Expected exclusion limits for Dark Matter annihilation. | 42 |
| 2.1 | Schematic picture of an electromagnetic cascade. | 49 |
| 2.2 | Schematic picture of a hadronic cascade. | 50 |
| 2.3 | $\gamma$-ray shower vs. proton shower. | 51 |
| 2.4 | Cherenkov light spectrum. | 52 |
| 2.5 | IACT principle. | 53 |
| 2.6 | Cherenkov light density vs. impact parameter. | 53 |
| 2.7 | Picture of the MAGIC telescope. | 54 |
| 2.8 | Technical drawing of the MAGIC telescope. | 56 |
| 2.9 | Schematic picture of the MAGIC PMT camera. | 57 |
| 2.10 | Schematics of the trigger system of the MAGIC telescope. | 58 |
| 2.11 | Current MAGIC read-out scheme. | 59 |
| 2.12 | Components of the MAGIC calibration system. | 61 |
| 2.13 | Status of MAGIC II in spring 2006. | 64 |
| 3.1 | Raw pulse shape recorded with the MAGIC FADCs. | 68 |
| 3.2 | Reconstructed high gain pulse shape. | 68 |
| 3.3 | Reconstructed pulse shapes | 69 |
| 3.4 | Noise autocorrelation average all pixels. | 73 |
| 3.5 | Digital Filter Weights | 75 |
| 3.6 | Digital filter pulse fit. | 77 |
| 3.7 | Amplitude / timing resolution for MC pulses. | 78 |

# LIST OF FIGURES

| | | |
|---|---|---|
| 3.8 | Signal reconstruction bias for MC pulses. | 78 |
| 3.9 | Extracted pedestal distribution. | 79 |
| 3.10 | Pedestal RMS. | 79 |
| 3.11 | Timing resolution for LED calibration pulses. | 81 |
| 3.12 | Calibration results. | 83 |
| 3.13 | Shower picture before and after image cleaning. | 86 |
| 3.14 | Definition of the Hillas Parameters. | 87 |
| 3.15 | $\gamma$-ray shower vs. proton shower image. | 88 |
| 3.16 | Comparison of image parameters for MC $\gamma$-rays and OFF data. | 89 |
| 3.17 | RF Gini index. | 90 |
| 3.18 | Hadronness distribution for MC $\gamma$-rays and background. | 91 |
| 3.19 | RF energy estimation and energy resolution. | 92 |
| 3.20 | The Disp method to reconstruct the primary arrival direction. | 93 |
| 3.21 | Alpha plot. | 95 |
| 3.22 | Raw Disp maps in camera and sky coordinates. | 96 |
| 3.23 | Camera acceptance map, background model. | 96 |
| 3.24 | Background subtracted sky map. | 97 |
| 3.25 | $\theta^2$ plot. | 99 |
| 3.26 | Pointing Deviation as determined with the MAGIC starguider. | 100 |
| 3.27 | Interpolated and folded sky maps. | 101 |
| 3.28 | Effective area as function of estimated and true energy. | 102 |
| 3.29 | Effective observation time: distribution of time differences. | 104 |
| 3.30 | Distribution of $\gamma$-ray excess events vs. $E_{est}$ and $E_{true}$. | 105 |
| 3.31 | Energy spectrum of the Crab Nebula. | 106 |
| 3.32 | Source location in the camera for wobble mode observations. | 107 |
| 3.33 | Camera acceptance in wobble mode. | 108 |
| 3.34 | Relative effective area wobble/ON vs. energy. | 109 |
| 3.35 | Alpha angle definition for wobble data. | 110 |
| 3.36 | Alpha plot for wobble data. | 110 |
| 3.37 | Computation of the background map for wobble data. | 111 |
| 3.38 | Raw wobble Disp map, definition of ON/OFF samples. | 112 |
| 3.39 | $\theta^2$ plot for wobble data. | 113 |
| 3.40 | MAGIC flux sensitivity. | 115 |
| 3.41 | Angular resolution. | 116 |
| 3.42 | Comparison of image parameters for MC $\gamma$-rays and $\gamma$-ray candidates. | 117 |
| 3.43 | Large zenith angle observations. | 119 |
| 3.44 | Collection area and threshold vs. ZA. | 119 |
| 3.45 | Energy spectrum of the Crab Nebula. | 120 |
| 3.46 | Collection area and threshold vs. ZA. | 121 |
| 4.1 | The star field around the Galactic Center. | 126 |
| 4.2 | Sky map of the Galactic Center. | 128 |
| 4.3 | $\theta^2$ plot for the Galactic Center. | 129 |
| 4.4 | $\theta^2$ plot for the composite SNR G 0.9+0.1. | 130 |
| 4.5 | VHE $\gamma$-ray energy spectrum of the Galactic Center. | 130 |

## LIST OF FIGURES

| | | |
|---|---|---|
| 4.6 | Light curve in VHE $\gamma$-rays of the Galactic Center. | 131 |
| 4.7 | The star field around HESS J1813-178. | 138 |
| 4.8 | Sky map of HESS J1813-178. | 139 |
| 4.9 | $\theta^2$ plot for HESS J1813-178. | 140 |
| 4.10 | VHE $\gamma$-ray energy spectrum of HESS J1813-178. | 141 |
| 4.11 | Leptonic and hadronic models for HESS J1813-178. | 143 |
| 4.12 | The star field around HESS J1834-087. | 149 |
| 4.13 | Sky map of HESS J1834-087. | 151 |
| 4.14 | Morphology of HESS J1834-087 above three different lower cuts in Size. | 152 |
| 4.15 | $\theta^2$ plot for HESS J1834-087. | 154 |
| 4.16 | VHE $\gamma$-ray energy spectrum of HESS J1834-087. | 154 |
| 5.1 | Muon rejection by timing. | 161 |
| 5.2 | Differences in timing structure for $\gamma$-rays and hadrons. | 162 |
| 5.3 | The multiplexed FADC concept. | 163 |
| 5.4 | Two channel fiber-optic delay module. | 164 |
| 5.5 | Four channel fiber-optic splitter module. | 165 |
| 5.6 | The printed circuit board for analog signal multiplexing. | 166 |
| 5.7 | Circuit diagram of the Digital Switch Control circuit (DSC). | 167 |
| 5.8 | The printed circuit board for fast switching. | 168 |
| 5.9 | Schematics of the fast switches. | 169 |
| 5.10 | Schematics of the summing stage. | 170 |
| 5.11 | Oscilloscope snapshot of two consecutive multiplexed signals. | 172 |
| 5.12 | Linearity of the switches. | 173 |
| 5.13 | Integration of the MUX-FADC read-out in the current MAGIC read-out. | 175 |
| 5.14 | Pedestals: Raw data. | 177 |
| 5.15 | Calibration: Raw data. | 177 |
| 5.16 | Mean reconstructed number of photoelectrons for the LED pulser. | 178 |
| 5.17 | Timing resolution determined for the LED pulser. | 179 |
| 5.18 | Pulse shapes for cosmic events. | 180 |
| 5.19 | Reconstructed signal and arrival times for a typical event. | 180 |
| 5.20 | Noise autocorrelation. | 181 |
| 5.21 | Integrated noise after calibration into photoelectrons. | 182 |
| 5.22 | Time structure in a pedestal event. | 182 |
| 5.23 | Time distribution of the structures on the pedestal baseline. | 183 |
| 5.24 | Reconstructed single photoelectron spectrum of a MC pedestal run. | 184 |
| 5.25 | Comparison of the pulse arrival time resolution. | 185 |
| 5.26 | MUX QC at MPI. | 187 |
| 5.27 | MUX installation at La Palma | 188 |

# List of Tables

1.1 Properties of the Galactic Center. . . . . . . . . . . . . . . . . . . . . . . . 31
4.1 Data set per observation period of the GC. . . . . . . . . . . . . . . . . . . 127
5.1 Specifications of the electronics for analog signal multiplexing. . . . . . . . 165
5.2 Specifications of the ultra-fast FADC. . . . . . . . . . . . . . . . . . . . . 171
5.3 Overview of the data taken during the MUX-FADC prototype test. . . . . 176

# Bibliography

Aharonian, F. A., 2001, SSRv, 99, 187.

Aharonian, F. A., 2004, Very High Energy Cosmic Gamma Radiation, River Edge, NJ: World Scientific Publishing.

Aharonian, F. A. et al., 1994, J. Phys. G., 21, 419.

Aharonian, F. A. et al., 2001, Astropart. Phys., 15, 335.

Aharonian, F. A. et al. (HEGRA Collab.), 2002, A&A, 395, 803.

Aharonian, F. A. et al. (HEGRA Collab.), 2003, A&A, 403, L1.

Aharonian, F. A. et al. (HEGRA Collab.), 2004a, A&A, 614, 897.

Aharonian, F. A. et al. (HESS Collab.), 2004b, A&A, 425, L13.

Aharonian, F. A. et al. (HESS Collab.), 2004c, Nature, 432, 75.

Aharonian, F. A. et al. (HESS Collab.), 2005a, Science, 307, 1938.

Aharonian, F. A. et al. (HESS Collab.), 2005b, A&A, 432, L25.

Aharonian, F. A. et al. (HESS Collab.), 2005c, A&A, 437, L7.

Aharonian, F. A. et al. (HESS Collab.), 2005d, A&A, 442, 1.

Aharonian, F. A. et al. (HESS Collab.), 2005e, Science 309, 746.

Aharonian, F. A. et al. (HESS Collab.), 2006a, ApJ, 636, 777.

Aharonian, F. A., et al. (HESS Collab.), 2006b, Nature, 440, 1018.

Aharonian, F. A., et al. (HESS Collab.), 2006c, Nature, 439, 695.

Aharonian, F. A., et al. (HESS Collab.), 2006d, A&A, 457, 899.

Aharonian, F. A., Atoyan, A. M. & Kifune, T., 1997, MNRAS, 291, 162.

Aharonian, F. A. & Atoyan, A. M., 2000, A&A, 362, 937.

Aharonian, F. A., Drury, L. O'C & Voelk, H. J., 1994, A&A, 285, 645.

Aharonian, F. A. & Neronov, A., 2005, ApJ, 619, 306.

Ahrens, J. et al. (IceCube Collab.), 2004, Astropart. Phys., 20, 507.

Albert, J. et al. (MAGIC Collab.), 2006a, ApJ, 637, L41.

Albert, J. et al. (MAGIC Collab.), 2006b, ApJ, 638, L101.

Albert, J. et al. (MAGIC Collab.), 2006c, ApJ, 643, L53.

Albert, J. et al. (MAGIC Collab.), 2006c, ApJ, 639, 761.

Albert, J. et al. (MAGIC Collab.), 2006d, Science 312, 1771.

Albert, J. et al. (MAGIC Collab.), 2006e, submitted to NIM A, astro-ph/0612385.

Aliu, E. & Wittek, W., 2006, MAGIC-TDAS 06-01.

Alcaraz, J. et al., 2000a, Phys. Lett., 490, 27.

Alcaraz, J. et al., 2000b, Phys. Lett., 494, 392.

Allen, G. E., Petre, R. & Gotthelf, E. V., 2001, ApJ, 558, 739.

Amenomori, M. et al, 2002, ApJ, 580, 887.

Antoranz, P. et al. (MAGIC Collab.), 2006, Proc. of the ECRC 2006 Lisbon, MPP-2006-143.

Anykeev, V. B., Spiridonov, A. A. & Zhigunov, V.B., 1991, NIM, A303, 350.

Ariskin V. I. & Berulis I. I., 1970, Soviet Astronomy, 13, 883.

Arons, J., 1998, Mem. Scoc. Ast. Ital., 69, 989.

Atkins, R. et al., 2004, ApJ, 608, 680.

Atkins, R. et al., 2005, Phys. Rev. Lett., 95, 1103.

Atoyan, A. & Dermer, C. D., 2004, ApJ, 617, L123.

Baade, W. & Zwicky, F., 1934, Phys. Rev., 46, 76.

Baganoff, F. K., 2001, Nature, 413, 45.

Baganoff, F. K., 2003, ApJ, 591, 891.

Baixeras, C. et al. (MAGIC Collab.), 2004, NIM, A518, 188.

Barrio, J. A. et al. (MAGIC Collab.), *The MAGIC Telescope - Design Study for the Construction of a 17m Cherenkov Telescope for Gamma Astronomy above 10 GeV*, MPI-PhE-98-05.

# BIBLIOGRAPHY

Bartko, H. et al. (MAGIC Collab.), 2005a, Proceedings of the conference *Towards a Network of Atmospheric Cherenkov Detectors VII*, Palaiseau, France, astro-ph/0506459.

Bartko, H. et al., 2005b, NIM, A548, 464.

Bartko, H. et al. (MAGIC Collab.), Proc. of the 29th ICRC, Pune, India, 4-17, astro-ph/0508244.

Bastieri, D., 2005, Astropart. Phys., 23, 572.

Battaglia, G. et al. 2005, MNRAS, 364, 433.

Bednarek, W., 1997, MNRAS, 285, 69.

Bednarek, W., 2002, MNRAS, 331, 483.

Bednarek, W., 2006, Ap&SS in press, astro-ph/0610307.

Bell, A. R., 1978, MNRAS, 182, 147.

Berezhko, E. G. & Völk, H. J., 1997, Astropart. Phys., 7, 183.

Berezhko, E. G. & Völk, H. J., 2000, Astropart. Phys., 14, 201.

Bergstrom, L., Ullio, P. & Buckley, J. H., 1998, Astropart. Phys., 9, 137.

Bergstrom, L. et al., 2004, Phys. Rev. Lett., 94, 1301.

Bernlöhr, K., 2000, Astropart. Phys., 12, 255.

Bertone, G., Servant, G., & Sigl, G., 2003, Phys. Rev., D68, 044008.

Bethe, H. A., 1990, Rev. Mod. Phys., 62, 801.

Biermann, P. L. et al., 2004, ApJ, 604, L29.

Bigongiari, C. et al., 2004, NIM, A518, 193.

Biller, S. D. et al., 1995, Astropart. Phys., 3, 385.

Biller, S. D. et al., 1999, Phys. Rev. Lett., 83, 2108.

Bird, A. J., 2006, ApJ, 636, 765.

Blach, O & Martinez,M., 2005, Astropart. Phys., 23, 598.

Blatning, S. R. et al., 2000, NASA/TP-2000-210640.

Blanch, O., 2003, MAGIC-TDAS 03-06.

Blandford, R. & Eichler, D., 1987, Phys. Rep., 150, 1.

Blumenthal, G. R. & Gould, R. J., 1970, Rev. Mod. Phys., 42, 237.

Bock, R. K. et al., 2004, NIM, A516, 511.

Bossa, M., Mollerach, S. & Roulet, E., 2003, J. Phys. G, 29, 1409.

Bower, C. B. et al., 2004, Science, 304, 704.

Breiman, L., 2001, Machine Learning, 45, 5.

Bretz, T. et al. (MAGIC Collab.), 2003, Proc. of the 28th ICRC, Tsukuba, Japan, 2943.

Bretz, T. et al. (MAGIC Collab.), 2005, Proc. of the 29th ICRC, Pune, India, 4-311, astro-ph/0508274.

Bretz, T. & R. Wagner (MAGIC Collab.), 2003, Proc. of the 28th ICRC, Tsukuba, Japan, 2947.

Brogan, C. L. et al., 2004, ApJ, 127, 335.

Brogan, C. L. et al., 2005, ApJ, 629, L105.

Camilo, F. et al., 2002, ApJ, 579, L25.

Chitnis, V. R. & Bhat, P. N., 2001, Astropart. Phys., 15, 29.

Churchwell, E., 1990, AApR, 2, 79.

Cleland, W. E. & Stern, E. G., 1994, NIM, A338, 467.

Clemens, D. P., 1985, ApJ, 295, 422.

Coil, A. L. & Ho, P. T. P., 2000, ApJ, 533, 245.

Cojocaru, C. et al. (ATLAS Liquid Argon Collab.), 2004, NIM, A531, 481.

Colin, P., 2006, Proceedings of TeV particle astrophysics 2, Madison, August 2006, astro-ph/0610344.

Commichau, S. 2004, private communication.

Condon, J. J. et al., 1998, AJ, 115, 1693.

Cordes, J. M. & Lazio, T. J. W., 2002, astro-ph/0207156.

Coroniti, F. V., 1990, ApJ, 349, 538.

Cortina, J., 2005, MAGIC-TDAS 05-04.

Cortina, J. et al. (MAGIC Collab.), 2003, Proc. of the 28th ICRC, Tsukuba, Japan, 2931.

Cortina, J. et al. (MAGIC Collab.), 2005, Proc. of the 29th ICRC, Pune, India, 5-359, astro-ph/0508274.

Crocker, R.M. et al., 2005, ApJ, 622, 892.

Cronin, J., Gaisser, T. K. & Swordy, S. P., 1997, Sci. Amer., 276, 44.

Dame, T. M., Elmegreen, B. G., Cohen, R. S. & Thaddeus, P., 1986, ApJ, 305, 892.

Dame, T. M., Hartmann, D. & Thaddeus, P., 2001, ApJ, 547, 792.

Daum, A. et al. (HEGRA Collab.), 1997, Astropart. Phys., 8, 1.

Dermer, C. D., 1986, A&A, 157, 223.

Dickey, J. M. & Lockman, F. J., 1990, ARA&A, 28, 215.

Diehl, R. 1988, SSRv, 49, 85.

Domingo-Santamaria, E & Torres, D. F., 2005, A&A, 444, 403.

Domingo-Santamaria, E. et al. (MAGIC Collab.), 2005, Proc. of the 29th ICRC, Pune, India, 5-363, astro-ph/0508274.

Dwek, E. & Krennrich, F., 2005, ApJ, 618, 657.

Dyson, J. 1997, Physics of the Interstellar Medium, London: Taylor & Francis.

Eidelman, S. et al. (Particle Data Group), 2004, Physics Letters, B592, 1.

Eisenhauer, F., et al., 2003, ApJ, 597, L121.

Eisenhauer, F., et al., 2005, ApJ, 628, 246.

Ellis, J. R. et al. 1984, Nucl. Phys., B238, 453.

Ellison, D. C. & Reynolds, S. P., 1991, ApJ, 382, 242.

Elsässer, D. & Mannheim, K., 2005, Phys. Rev. Lett., 94, 171302.

Elterman, L., 1964, Appl. Opt., 3, 6.

Enomoto, R. et al (CANGAROO Collab.). 2002, Nature, 416, 823.

Enomoto, R. et al (CANGAROO Collab.). 2006, ApJ in press, astro-ph/0608422.

Evans, N. W. et al., 2004, Phys.Rev., D69, 123501.

Falcke, H., Mannheim, K., Biermann, P.L., 1993, A&A, 278, L1.

Falcke, H. & Markoff, S., 2000, A&A, 362, 113.

Fatuzzo, M. & Melia, F., 2003, ApJ, 596, 1035.

Fazio, G. G. & Stecker, F. W., 1970, Nature, 226, 135.

Fegan, D. J., 1997, J. Phys. G., 23, 1013.

Fermi, E., 1949, Phys. Rev., 75, 1169.

Fermi, E., 1954, ApJ, 119, 1.

Fernow, R. C., 1986, Introduction to Experimental Particle Physics, Cambridge University Press.

Firpo, R., 2006, Ph.D. thesis IFAE Barcelona.

Fleysher, R. et al. (MILAGRO Collab.), 2005, AIP Conf. Proc., 745, 269.

Flix, J., 2005, Proc. of Rencontres de Moriond, La Thuile, Italy, astro-ph/0505313.

Flix, J., 2006, Ph.D. thesis IFAE Barcelona.

Fomin, V. P. et al., 1994, Astropart. Phys., 2, 137.

Fornengo, N. et al., 2004, Phys. Rev., D70, 103529.

Fossati, G., et al., 1998, MNRAS, 299, 433.

Gaensler, B. M. & Johnston, S., 1995, MNRAS, 275, L73.

Gaensler, B. M., 2000, MNRAS, 317, 58.

Gaisser, T. K., 1990, Cosmic rays and particle physics, Cambridge University Press.

Gaisser, T. K., AIP Conf. Proc., 558, 27.

Garcia-Munoz, M., 1987, ApJS, 64, 269.

Garczarczyk, M. et al. (MAGIC Collab.), 2003, Proc. of the 28th ICRC, Tsukuba, Japan, 2935.

Garczarczyk, M., 2006, Ph.D. thesis, University of Rostock.

Garczarczyk, M. et al., 2006, MAGIC-TDAS 06-05.

Gascon, J. 2005, astro-ph/0504241.

Gaug, M. et al. 2004, MAGIC-TDAS 04-06.

Gaug, M., Bartko, H. et al. (MAGIC Collab.), 2005, Proc. of the 29th ICRC, Pune, India, 5-375, astro-ph/0508274.

Gaug, M., 2006, Ph.D. thesis IFAE Barcelona.

Gebhardt, K et al., 2000, ApJ, 539, L13.

Genzel, R., et al., 2003, Nature, 425, 934.

Ghez, A. M., 2000, Nature, 407, 349.

Ghez, A. M., 2003a, ApJ, 586, L127.

Ghez, A. M., 2003b, ANS, 324, 527.

Ginzburg, V. L. & Syrovatskii, S. I., 1964, The Origin of Cosmic Rays, New York: Macmillan.

Gnedin, O. et al., 2004, ApJ., 616, 16.

Goebel, F. et al. (MAGIC Collab.), 2003, Proc. of the 28th ICRC, Tsukuba, Japan, 2003.

Goebel, F. et al. (MAGIC Collab.), 2005, Proc. of the 29th ICRC, Pune, India, 5-375, astro-ph/0508274.

Goldreich, P. & Julian, W. H., 1969, ApJ, 157, 869.

Goodman, J. A. et al., 2006, Procs. 2nd Workshop On TeV Particle Astrophysics, Madison, WI, USA.

Gordon, S. T. & Gordon, M. A., 1970, ApJ, 162, L93.

Gould, R. J. & Schreder, G. 1966, Phys. Rev. Lett., 16, 252.

Grasso, D. & Maccione, L., 2005, Astropart. Phys., 24, 273.

Green, D. A., 2004, BASI, 32, 335.

Greisen, K., 1966, Phys. Rev. Lett., 16, 748.

C. Grupen, A. Bohrer, and L. Smolik, 1996, Cambridge Monogr. Part. Phys. Nucl. Phys. Cosmol., 5, 1.

Halzen, F. & Hooper, D., 2002, Rept. Prog. Phys, 65, 1025.

Han, J. L. & Tian, W. W., 1999, A&AS, 136, 571.

Handa, T. et al., 1987, PASJ, 39, 709.

Harding, A. K., 1981, ApJ, 245, 267.

Harding, A. K., 1996, Space Sci. Rev., 75, 257.

Hartman, R. C. et al., 1999, ApJS, 123, 79.

Haslam, C. G. T., 1982, A&AS, 47, 1.

Haungs, A., 2003, J. Phys. G., 29, 809.

Hauser, M. G.; Dwek, E., 2001, ARA&A, 39, 249.

Hayashida, N. et al. (AGASA Collab.), 1999, Astropart. Phys., 10, 303.

Hayashida, M. et al. (MAGIC Collab.), 2005, Proc. of the 29th ICRC, Pune, India, 5-183, astro-ph/0508274.

Haynes, R. F., Caswell, J. L., & Simons, L. W. J. 1978, Austr. J. of Phys. Astrophys. Suppl., 45, 1

Heck, D. et al., 1998, Report FZKA 6019.

Heinz, S. and Sunyaev, R. A., 2002, A&A, 390.

Heitler, W., 1954, The Quantum Theory of Radiation, Oxford, Clarendon Press.

Helfand, D. J., Becker, R. H. and White, R. L., 2005, astro-ph/0505392.

Hengstebeck, T., 2003, http://magic.physik.hu-berlin.de/protected/ranforest/ .

Hess, V., 1912, Physik. Zeitschr., 13, 1084.

Hillas, A. M., 1985, Proc. of the 19th ICRC, La Jolla, 3, 445.

Hillas, A. M., 1996, SSRv, 75, 17.

Hillas, A. M., 2005, J. Phys., G31, 95.

Hillas, A. M. (Whipple Collab.), 1998, ApJ, 503, 744.

Hinton, J. A. & Aharonian, F. A., 2006, ApJ in press, astro-ph/0607557.

Holder, J. et al. (Veritas Collab.), 2006, Astropart. Phys., 25, 391.

Hooper, D. & Dingus, B., 2005, AdSpR, 35, 130, astro-ph/0212509.

Hooper, D. et al., 2004, JCAPP, 9, 2.

Horns, D., 2005, Phys. Lett., B607, 225.

Hurt, R., 2005, SSC/Caltech.

Jackson, J. D., 1975, Classical Electrodynamics, Wiley, New York.

Jackson, J. M., et al. 2006, ApJS, 163, 145.

de Jager, O. C.; Harding, A. K., 1992, ApJ, 396, 161.

Jelley, J. V., 1966, Phys. Rev. Lett., 16, 479.

Johnson, W. N., 1993, A&AS, 97, 21.

Jones, T. W., et al., 1998, PASP, 110, 125.

Jones, F. C. & Ellison, D. C., 1991, Space Sci. Rev., 58, 259.

Jun, B.-I. & Jones, T. W., 1999, ApJ, 511, 774.

Jungman, G., Kamionkowski, M. & Griest, K., 1996, Phys. Rep., 267, 195.

Kabuki, S. et al. (CANGAROO Collab.), 2003, NIM, A500, 318.

Kamae, T., Abe, T. & Koi, T., 2005, ApJ, 620, 244.

Kampert, K.-H. (AUGER Collab.), 2006, astro-ph/0608136.

Kanbach, G., 1988, SSRv, 49, 61.

Kappes, A. et al, 2006, astro-ph/0607286.

Karle, A. et al., 1995, Astropart. Phys., 3, 321.

Kashlinsky, A., 2005, PhR, 409, 361.

Kassim, N. E., 1992, AJ, 103, 943.

Katagiri, H. (CANGAROO Collab.), 2005, ApJ, 619, L163.

Katz, U., 2006, NIM, A567, 457.

Kembhavi, A. K. & Narlikar, J. V., 1999, "Quasars and active galactic nuclei", Cambridge University Press, Cambridge.

K. Kleinknecht, 1998, Cambridge Univ. Press, 2nd edition.

Klypin, A., Zhao, H. & Somerville, R. S., 2002, ApJ, 573, 597.

Kneizys et al., 1996, The MODTRAN 2/3 Report and LOWTRAN 7 Model, Phillips Laboratory, Hanscom AFB, MA 01731, USA.

Konopelko, A. et al., 1999, J Phys, G25, 1989.

Kosack, K. et al., 2004, ApJ, 608, L97.

Koyama, K., 1995, Nature, 378, 255.

Krawczynski, H. et al. (Whipple Collab.), 2004, ApJ, 601, 151.

Krennrich, F. et al. (Whipple Collab.), 1999, ApJ, 511, 149.

Landi, R. et al., 2006, ApJ, 651, 190.

LaRosa, T. N. et al., 2000, AJ, 119, 207.

LaRosa, T. N. et al., 2005, ApJ, 626, L23.

Lazendic, J. S. et al., 2004, ApJ, 602, 271.

T. J. W. Lazio, et al., 2006, available at http://rsd-www.nrl.navy.mil/7213/lazio/GC/GCSchematicscan2rot.gif.

Leo, W. R., 1994, Techniques for Nuclear and Particle Physics Experiments, Springer.

Lessard, R. W. et al. (Whipple Collab.), 2001, Astropart. Phys., 15, 1.

Letessier-Selvon, A. et al. (AUGER Collab.), , Proc. of the 29th ICRC, Pune, India, 7-67, astro-ph/0610160.

Li, T.-P. & Ma, Y.-Q., 1983, ApJ, 272, 317.

Lipson, J. & Harvey, G., 1983, Journal of Lightwave Technology, 1, 387.

Liu, S. et al., 2006, ApJ, 647, 1099.

Lokas, E. et al., 2005, MNRAS, L48.

Longair, M. S., 1994, "High energy astrophysics", Vol. 2, Cambridge university press.

Lorenz, E. et al., 2001, NIM, A461, 517.

Lorenz, E., 2003, Nucl. Phys. B (Proc.Suppl.), 114, 217.

Lovelace, R. V. E., 1976, Nature, 262, 649.

Lu, Y., Cheng, K. S. & Huang, Y. F., 2006, ApJ, 641, 288.

Lucarelli, F. et al. (HEGRA Collab.), 2003, Astropart. Phys., 19, 339.

Lyubarsky, Y. & Kirk, J.G., 2001, ApJ, 547, 437.

Maeda, Y. et al., 2002, ApJ, 570, 671.

Majumdar, P. et al. (MAGIC Collab.), 2005, Proc. of the 29th ICRC, Pune, India, 5-203, astro-ph/0508274.

Malkov, M. A., O'C Drury, L. , 2001, Rep. Prog. Phys., 64, 429.

Manchester, R. N., Hobbs, G. B., Teoh, A. & Hobbs, M. 2005, AJ, 129, 1993.

Mayer-Hasselwander, H. A. et al., 1998, A&A, 335, 161.

Mazin, D. et al., In preparation.

McKee, C. F. & Ostriker, J. P., 1977, ApJ, 218, 148.

McLaughlin, D. E., 1999, ApJ, 512, L9.

Melia, F. & Falcke, H., 2001, ARA&A, 39, 309.

Merck, Ch., 2004, Master thesis, University Siegen.

Meucci, M. et al. (MAGIC Collab.), 2004, NIM, A518, 554.

Mezger, P. et al., 1989, A&A, 209, 337.

Michel, F. C., 1982, Rev. Mod. Phys., 54, 1.

Mirabel, I. F. et al., 1992, Nature, 358, 215.

Mirabel, I. F., 2006, Science, 312, 1759.

Mirabel, I. F. & Rodriguez, L., 1994, Nature, 371, 46.

Mirabel, I. F. & Rodriguez, L., 1999, ARA&A, 37, 409.

Mirzoyan, R. et al., 2002, IEEE Trans. Nucl. Sci., 49, 2473 .

Mirzoyan, R. & Lorenz, E. 1997, Proc. of the 25th ICRC, Durban, South Africa, 7, 265.

Mirzoyan, R., Cortina, J. and Lorenz, E., 2001, Proc. of the 27th ICRC, Hamburg, Germany.

Mirzoyan, R. et al., 2006, Astropart. Phys., 25, 342.

Mizobuchi, S. et al. (MAGIC Collab.), 2005, Proc. of the 29th ICRC, Pune, India, 5-323, astro-ph/0508274.

Moore, B. et al., 1998, ApJ, 499, L5.

Moralejo, A., 2002, MAGIC-TDAS 02-05.

Morris, M. & Serabyn, E., 1996, ARA&A, 34, 645.

Moskalenko, I. & Strong, A. W., 1998, ApJ, 293, 694.

Nakamura et al., 1995, Jap. Journ. of Appl. Phys., 34, L1332.

Navarro, J., Frenk, C. & White, S. 1997, ApJ, 490, 493.

Nomoto, K., Thielemann, F.-K., Yokoi, K., 1984, ApJ, 286, 644.

Nyquist, H., 1928, Trans. AIEE, 47, 617.

Ostankov, A. et al., 2000, NIM, A442, 117.

Paneque, D. et al. (MAGIC Collab.), 2003, Proc. of the 28th ICRC, Tsukuba, Japan, 2927.

Paneque, D. et al., 2004, NIM, A 518, 619.

Papoulis, A., 1977, "Signal analysis", McGraw-Hill.

Paredes, J. M., 2005, AIPC, 745, 93.

Pohl, M., 1997, A&A, 317, 441.

Pohl, M., 2005, ApJ, 626, 174.

Porquet, D. et al., 2003, A&A, 407, L17.

Prada, F. et al., 2004, Phys. Rev. Lett., 93, 241301.

Press, W. H. et al., 2002, Cambridge University Press, second edition.

Punch, M. et al. (Whipple Collab.), 1992, Nature, 358, 477.

Quataert, E. and Loeb, A. 2005, ApJ, 635, L45.

Rees, M. J. & Sciama, D. W., 1966, Nature, 211, 805.

Rees, M. J., 1967, MNRAS, 137, 429.

Rees, M. J., 1984, ARA&A, 22, 471.

Reich, W. et al., 1990, A&AS, 85, 633.

Reid, M. J., 1993, ARA&A, 31, 345.

Richstone, D et al., 1998, Nature, 395, A14.

Riegel, B. et al. (MAGIC Collab.), 2005, Proc. of the 29th ICRC, Pune, India, 5-219, astro-ph/0508274.

Rose, J., 1995, Proc. "Towards a major atmospheric Cherenkov detector IV", Padova, Italy, 241.

Rossi, B., 1965, High-Energy Particles, Prentice Hall Series.

Rowell, G. P., 2003, A&A, 410, 398.

Rudak, B., 2001, astro-ph/0101138.

Ryan, J. M., 2004, SPIE, 5488, 977.

Safi-Harb, S., 2004, BAAA, 47, 277.

Scherrer, R. & Turner, M. 1986, Phys. Rev., D33, 1585.

Schlickeiser, R., 2002, "Cosmic ray astrophysics", Springer-Verlag, New York.

Schmidt, F., http://www.ast.leeds.ac.uk/ fs/showerimages.html

Schödel, R. et al., 2002, Nature, 419, 694.

Schweizer, T. et al., 2002, IEEE Trans. Nucl. Sci., 49, 2497.

Scott, W. T., 1963, Rev. Mod. Phys., 35, 231.

Seo, E. S. & Ptuskin, V. S., 1994, ApJ, 431, 705.

Shaver P. A. & Goss W. M., 1970, AuJPA 14, 133.

Shoup, A., 1994, AAS, 185, 1604.

Smith, A., 2005, Proceedings of the conference *Towards a Network of Atmospheric Cherenkov Detectors VII*, Palaiseau, France, 429.

Snowden, S. L. et al., 1997, ApJ, 485, 125.

Sobczynska, D., 2002, MAGIC-TDAS 02-10.

Stecker, F. W., 1971, Cosmic Gamma Rays (Baltimore: Mono).

Stecker, F. W., de Jager, O. C., Salamon, M. H. 1992, ApJ, 390, L49.

Sugizaki, M., 2001, ApJS, 134, 77.

Swordy, S. P. et al., 1990, ApJ, 349, 625.

Tanimori, T. (CANGAROO Collab.), 1994, ApJ, 429, L61.

Taylor, J. H., Cordes, J. M., 1993, ApJ, 411, 674.

Teshima, M. et al. (MAGIC Collab.), 2005, Proc. of the 29th ICRC, Pune, India, 5-227, astro-ph/0508274.

Tonello, N., 2006, Ph.D. thesis, Techical University Munich.

Torres, D. F. et al., 2003, Phys. Rept., 382, 303.

Turini, N. et al. (MAGIC collaboration), 2005, Proceedings of the conference *Towards a Network of Atmospheric Cherenkov Detectors VII*, Palaiseau, France, 493.

Tsuchiya, K. et al. (CANGAROO Collab.), 2002, Proc. of "The Universe Viewed in Gamma Rays", Kashiwa, Chiba, Japan, 25-28 Sep 2002, p. 17.

Tsuchiya, K. et al. (CANGAROO Collab.), 2004, ApJ, 606, L115.

Ubertini, P. et al., 2005, ApJ, 629, L109.

Uchida, K. I.; Guesten, R. 1995, A&A, 298, 473.

Uchiyama, Y. et al., 2002, ApJ, 571, 866.

Urry, C. & Padovani, P., 1995, PASP, 107, 803.

Völk, H. J., Berezhko, E. G. & Ksenofontov, L. T., 2005, A&A, 433, 229.

Wagner, R. et al. (MAGIC Collab.), 2005, Proc. of the 29th ICRC, Pune, India, 4-163, astro-ph/0508244.

Wang, Q. D. et al., 2006, MNRAS, 367, 937.

Weekes, T. C. (Whipple Collab.), , 1998, ApJ, 503, 744.

Werner, K., 1993, Phys. Rept., 232, 87.

White, R. L., Becker, R. H., & Helfand, D. J., 2005, AJ, 130, 586.

Wigmans, R., 1987, NIM, A259, 389.

Willman, B. et al., 2005, ApJ, 626, L85.

Winkler, C., 2003, A&A, 411, L1.

Winston, R., 1970, J. Opt. Soc. Amer., 60, 245.

Wittek, W., 2001, MAGIC-TDAS 01-05.

Wittek, W., 2002a, MAGIC-TDAS 02-02.

Wittek, W., 2002b, MAGIC-TDAS 02-03.

Wittek, W., 2002c, MAGIC-TDAS 02-14.

Wittek, W., 2005a, MAGIC-TDAS 05-05.

Wittek, W., 2005b, MAGIC-TDAS 05-07.

Woltjer, L., , 1972, ARA&A, 10, 129.

Wood, K., 1995, AAS, 187, 7120.

Woosley, S. E. & Weaver, T. A., 1995 ApJS, 101, 181.

Yao, W.-M. et al., 2006, J. Phys., G33, 1.

Yusef-Zadeh, F. et al., 1999, ApJ, 527, 172.

Yusef-Zadeh et al., 2002, ApJ, 568, L121.

Zatsepin, G. T., Kuzmin, V. A., 1966, JETP Lett., 4, 78.

Zhang, J. L. & Yuan, Y. F., 1998, ApJ, 493, 826.

Ziltoun, R., 2001, ATLAS-LARG-2001-003.

# Curriculum Vitae

Hendrik Bartko

| | |
|---|---|
| 1977/11/01 | born in Zossen near Berlin, Germany |
| 1984 - 1991 | elementary school Blankenfelde |
| 1991 - 1997 | Fontane-Gymnasium Rangsdorf |
| 1997 - 1998 | civil service, care for the elderly, Blankenfelde |
| 1998 - 2000 | Humboldt University Berlin, study of physics |
| 2001 - 2002 | University of Illinois at Chicago, study of physics and research assistant, D0 experiment, Fermi National Accelerator Laboratory, Batavia |
| 2002 - 2003 | MPI für Physik, München and TU München, Diploma thesis, ATLAS experiment, Centre Européenne pour la Recherche Nucléaire, Geneva |
| 2004 - 2006 | MPI für Physik, München and LMU München, Ph.D. thesis, MAGIC telescope, Observatorio del Roque de los Muchachos, La Palma |
| 2001 - 2002 | UIC, teaching assistant, physics laboratory courses |
| 2003 - 2004 | LMU München, teaching assistant, theoretical mechanics |
| 2005 - 2006 | LMU München, teaching assistant, nuclear and particle physics |

The results of this theses have led to the following refereed publications:

*Tests of a Prototype Multiplexed Fiber-Optic Ultra-Fast FADC Data Acquisition System for the MAGIC Telescope*
H. Bartko et al., 2005, NIM, A548, 464.

*MAGIC Observations of Very High Energy Gamma-Rays from HESS J1813-178,*
J. Albert et al (MAGIC Collaboration), 2006, ApJ, 637, L41.

*Observation of Gamma-Rays From the Galactic Center with the MAGIC Telescope,*
J. Albert et al (MAGIC Collaboration), 2006, ApJ, 638, L101.

*Observation of VHE Gamma Radiation from HESS J1834-087/W41 with the MAGIC Telescope,*
J. Albert et al (MAGIC Collaboration), 2006, ApJ, 643, L53.

*Signal Reconstruction for the MAGIC Telescope*,
J. Albert et al (MAGIC Collaboration), 2006, NIM A submitted, astro-ph/0612385.

Papers presented at competitive international conferences:

*FADC Pulse Reconstruction Using a Digital Filter for the MAGIC Telescope*,
H. Bartko et al, Proceedings of 7th Workshop on Towards a Network of Atmospheric Cherenkov Detectors 2005, Palaiseau, France, 27-29 Apr 2005, astro-ph/0506459.

*Calibration of the MAGIC Telescope*,
M. Gaug, H. Bartko et al. (MAGIC collaboration), 29th International Cosmic Ray Conference (ICRC 2005), Pune, India, 3-11 Aug 2005, astro-ph/0508274.

*Toward Dark Matter Searches with the MAGIC Telescope*
H. Bartko et al (MAGIC collaboration), 29th International Cosmic Ray Conference (ICRC 2005), Pune, India, 3-11 Aug 2005, astro-ph/0508273.

*Observation of VHE Gamma Radiation from HESS J1834-178 with the MAGIC Telescope*,
H. Bartko et al (MAGIC collaboration), 29th International Cosmic Ray Conference (ICRC 2005), Pune, India, 3-11 Aug 2005, astro-ph/0508244.

*Search for Gamma Rays from the Galactic Center with the MAGIC Telescope*,
H. Bartko et al (MAGIC collaboration), 29th International Cosmic Ray Conference (ICRC 2005), Pune, India, 3-11 Aug 2005, astro-ph/0508244.

Public Outreach:

*Gamma Astronomie mit dem MAGIC Teleskop*,
H. Bartko et al, 2006, year book of the Max-Planck society,
http://www.mpg.de/bilderBerichteDokumente/dokumentation/jahrbuch/2006/physik/forschungsSchwerpunkt/index.html.

*The MAGIC Telescope Project for Observing Very High Energy Gamma-ray Cosmic Sources*,
H. Bartko et al, 2006, annual report of the Max-Planck Institute for Physics, Munich.

Invited talks:

2005 University of Würzburg: *Observation of Galactic Sources at Large Zenith Angle.*

2006 ISAF Rome: *Status, Results and Prospects of the MAGIC telescope.*

2006 MPI für Kernphysik, Heidelberg: *Observation of Galactic Sources with the MAGIC Telescope.*